The New Earth Blueprint

A Quantum Genesis

MATTHEW HERBERTSON

The New Earth Blueprint : A Quantum Genesis

In an era defined by critical planetary crossroads, a vision emerges that transcends mere sustainability: *The New Earth Blueprint: A Quantum Genesis*. This groundbreaking work unveils the audacious plan to fundamentally reimagine our world, delving into the very fabric of existence to forge a truly new beginning. More than just a conceptual framework, this blueprint details a revolutionary approach to planetary design, rooted in the cutting-edge principles of quantum mechanics.

Prepare to journey into a future where advanced scientific understanding meets audacious ambition. Discover how working at the most fundamental levels can unlock possibilities for ecological restoration, societal harmony, and a vibrant existence unseen before. This is not merely an incremental step; it's a *genesis* – a deliberate act of creation, a complete re-engineering of our planet from its quantum core. With implications that reach farther than our planet alone.

At the heart of this quantum transformation lies an unprecedented promise: an "Everybody Wins" system. Envision a world where the shadows of scarcity, starvation, and homelessness vanish forever. Imagine an Earth cleansed of pollution, where pain and suffering become echoes of the past, and the grip of fear, violence, and even death are distant memories. This paradigm shift paves the way for a future where every belief system, every religion not only coexists but flourishes in perfect harmony. *The New Earth Blueprint: A Quantum Genesis,* is your guide to understanding the grand design for a future where humanity not only survives but thrives, building an idealised world born from profound understanding and visionary intent.

One we will build, Together...

Understanding the Ascension - A Quantum Leap in Consciousness

We stand at the threshold of a profound transformation, a monumental shift in consciousness that will fundamentally reshape our world and the very fabric of reality as we know it. This is the ascension, the grand transition from the dense, limited third-dimensional reality to the expansive, luminous higher frequencies of the New Earth. It is a quantum leap in consciousness, a recalibration of our collective vibration to align with the heart of the cosmos.

This shift is not a sudden, cataclysmic event, but rather an intricate, ongoing process, a gradual and deliberate awakening of humanity to its inherent, divine potential. It is a multi-layered unfolding, a journey of remembering who we truly are. This is a movement of epic proportions, a collective evolution from the constricting grip of fear to the liberating embrace of unconditional love, from the illusion of separation to the profound realisation of unity, from the limitations of perceived scarcity to the boundless possibilities of infinite abundance.

The old Earth, constructed upon the shaky foundations of duality, incessant competition, and the artificial construct of scarcity, is now undergoing a necessary dissolution. Its outdated structures, rigid systems, and limiting beliefs are crumbling under the weight of their own incongruence, making way for a vibrant, new paradigm. This emerging paradigm is built upon the pillars of genuine collaboration, radical compassion, and the inherent abundance that flows from a consciousness of unity.

This transition, while glorious in its potential, is not without its inherent challenges. As the old, familiar structures begin to break down and dissolve, we may find ourselves experiencing moments of apparent chaos, unsettling uncertainty, and the lingering echoes of fear. However, these experiences are not signs of failure, but rather necessary and integral stages of profound growth. They are opportunities, presented with divine timing, to consciously release what no longer serves our highest evolution, to shed the outdated skins of our former selves, and to wholeheartedly embrace the luminous potential of the new.

The New Earth is not a distant, unattainable utopia, but a tangible, evolving reality that we are actively co-creating in every sacred moment. It is a world where we live in harmonious resonance with the rhythms of nature, where we honour and celebrate the interconnectedness of all sentient beings, and where we fearlessly express the fullness of our divine potential. It is a world where intuition guides our steps, where compassion informs our actions, and where love is the guiding principle of our existence.

This transformative shift is not merely about changing the external world around us, but about a deep, internal metamorphosis, a radical transformation of ourselves from the inside out. It is about awakening to our true, eternal nature as spiritual beings, remembering our profound connection to the divine source, and embodying the radiant energy of love in every aspect of our lives, in every thought, word, and deed. It is a journey of self-discovery, a pilgrimage to the heart of our own being.

Therefore, embrace this sacred moment of profound transformation. Surrender to the natural flow of change, trusting in the divine orchestration of the process. For within this monumental shift lies the boundless potential for a world beyond our wildest dreams, a world where love reigns supreme, where harmony prevails, and where humanity flourishes in its full, radiant glory. It is a world where we remember our true selves, where we live in unity, and where we create a reality that reflects the highest aspirations of our souls.

Navigating the Shift – The Path Inward

An Introduction to the Inward Journey

The journey to align with the new Earth begins not with grand gestures or external pursuits, but with a profound inward exploration. It's a journey of self-discovery, a peeling away of layers of fear, doubt, and limiting beliefs that have kept us tethered to the old paradigm. This is where the true work lies – in the quiet spaces within, where we confront the shadows and embrace the light.

This inward journey requires courage and honesty. It's about facing the parts of ourselves we've hidden away, the wounds we've ignored, and the patterns that no longer serve us. It's about acknowledging our fears without judgment, and choosing, moment by moment, to align with love.

Think of it as recalibrating your inner compass. For too long, we've relied on external validation and societal conditioning to guide our path. Now, we must learn to listen to the whispers of our own hearts, the intuitive nudges of our souls.

This recalibration involves several key aspects. First, it's about cultivating self-awareness. Becoming mindful of our thoughts, emotions, and reactions allows us to identify the patterns that keep us stuck. Second, it's about practicing self-compassion. Treating ourselves with kindness and understanding, especially when we stumble, is essential for healing and growth.

Third, it's about cultivating inner stillness. In the quiet moments of meditation or contemplation, we can connect with our inner wisdom and receive guidance from our higher selves. Fourth, it's about practicing forgiveness – forgiving ourselves and others for past mistakes and releasing the burdens of resentment and anger. These lower vibrations, hold us in the lower dimensions.

As we embark on this inward journey, we begin to dismantle the illusions of the old Earth paradigm and awaken to the truth of our interconnectedness. We realise that we are not separate from the Earth or from each other, but integral parts of a unified whole.

Tools for Inner Work:

These tools, when practiced consistently, create a space for inner transformation.
They help us to:

Quiet the mental chatter: Allowing us to access deeper levels of intuition and inner wisdom.
Heal emotional wounds: Releasing past traumas and negative patterns that hold us back.
Cultivate self-love: Building a foundation of self-acceptance and appreciation.
Increase our vibrational frequency: Aligning ourselves with the higher energies of the new Earth.

Deepening Meditation Practices:

Meditation, at its core, is about cultivating awareness. It's a practice that allows us to step back from the constant stream of thoughts and emotions, creating space for stillness and clarity.

To deepen your meditation, consider exploring these techniques:

Guided Meditations: These can be particularly helpful for beginners or when you're seeking specific outcomes, such as stress reduction or emotional healing. Guided meditations use visualisation, affirmations, and soothing voices to lead you into a relaxed state.

Mantra Meditation: Repeating a word, phrase, or sound (a mantra) can help focus the mind and quiet the mental chatter. Choose a mantra that resonates with you, such as "peace," "love," or "I am."

Breath work Meditation: Focusing on the breath is a powerful way to anchor yourself in the present moment. Observe the natural rhythm of your breath, noticing the rise and fall of your abdomen. Deep, conscious breathing can also help regulate your nervous system and promote relaxation.

Walking Meditation: This combines mindfulness with physical movement. Pay attention to the sensations of your feet touching the ground, the rhythm of your steps, and the sights and sounds around you.

Addressing Common Challenges in Meditation:

It's natural to encounter challenges when meditating. Don't get discouraged if your mind wanders or if you experience restlessness. Here are some tips:

Start small: Even a few minutes of daily meditation can make a difference. Gradually increase the duration as you become more comfortable.

Find a comfortable posture: You can sit, lie down, or even walk. The key is to find a position that allows you to relax without falling asleep.

Don't judge your thoughts: When thoughts arise, simply acknowledge them and gently redirect your attention back to your chosen focus.

Be patient: Meditation is a practice, not a performance. It takes time and consistency to develop a calm and focused mind.

Integrating Meditation into Daily Life:

Meditation doesn't have to be confined to a specific time or place. You can incorporate mindfulness into your daily activities by:

Setting intentions: Before starting any task, take a moment to set an intention for how you want to approach it.

Taking mindful breaks: Throughout the day, pause for a few moments to breathe deeply and reconnect with your inner stillness.

Practicing gratitude: Before going to sleep, reflect on the things you're grateful for.

By deepening your meditation practice, you can cultivate inner peace, clarity, and connection with your higher self.

Expanding Mindfulness:

Mindfulness is the practice of paying attention to the present moment, without judgment. It's about being fully engaged in whatever you're doing, rather than being caught up in thoughts about the past or future. To expand your mindfulness practice, consider applying to a regular activity.

Applying Mindfulness to Various Activities:

Mindful Eating: Pay close attention to the taste, texture, and smell of your food. Notice the sensations in your body as you eat. Avoid distractions like phones or television.

Mindful Walking: As you walk, focus on the sensations of your feet on the ground, the movement of your body, and the sights and sounds around you.

Mindful Conversations: When talking to others, give them your full attention. Listen deeply, without interrupting or formulating your response.

Cultivating Mindful Awareness of Emotions and Thoughts: Observe your emotions and thoughts without judgment. Simply acknowledge them and allow them to pass. Notice the physical sensations that accompany your emotions.
Recognise that your thoughts are not facts, but simply mental events.

Using Mindfulness to Reduce Stress and Anxiety: When you feel stressed or anxious, take a few deep breaths and bring your attention to the present moment.

Focus on your senses: what you can see, hear, smell, taste, and touch. Practice body scan meditation, paying attention to the sensations in each part of your body.

Key Aspects of Expanding Mindfulness:

Non-Judgmental Observation: Being aware of your experiences without labelling them as good or bad.

Present Moment Focus: Directing your attention to what is happening right now, rather than dwelling on the past or worrying about the future.

Acceptance: Acknowledging your experiences as they are, without trying to change them.

Compassion: Treating yourself with kindness and understanding, especially when you are struggling.

By expanding your mindfulness practice, you can cultivate greater awareness, reduce stress, and enhance your overall well-being.

Advanced Forgiveness Techniques:

Forgiveness is a powerful act of self-liberation. It's not about condoning harmful actions, but about releasing the emotional burdens that keep us bound to the past. Here's how to deepen your forgiveness practice:

Understanding the Deeper Layers of Forgiveness: Forgiveness is a process, not a one-time event. It requires patience and self-compassion. True forgiveness is about releasing resentment and anger, not about forgetting or excusing harmful behaviour. Forgiveness benefits the forgiver more than the forgiven. It frees us from the grip of negative emotions.

Practicing Forgiveness for Oneself and Ancestral Lines:

Self-Forgiveness: Acknowledge your mistakes and learn from them. Treat yourself with kindness and understanding. Release self-blame and self-criticism.

Ancestral Forgiveness: Recognise that you may carry emotional patterns and traumas from your ancestors. Practice forgiveness for your lineage, releasing these burdens for yourself and future generations.

Releasing Karmic Patterns Through Forgiveness: Karmic patterns are recurring cycles of behaviour and experience. Forgiveness can help break these cycles.
Reflect on past relationships and situations where you experienced pain or conflict.
Visualise yourself releasing the emotional cords that connect you to these experiences.
Use affirmations such as "I forgive myself, and I forgive all others involved. I release all negative karmic bonds."

Key Aspects of Advanced Forgiveness:

Radical Acceptance: Acknowledging the reality of what happened, without resistance.

Empathy: Understanding the other person's perspective, even if you don't agree with their actions.

Compassion: Extending kindness and understanding to yourself and others.

Release: Letting go of the need for revenge or retribution.

By practicing advanced forgiveness techniques, you can heal deep emotional wounds, release karmic patterns, and create space for greater peace and freedom in your life.

Cultivating Profound Self-Compassion:

Self-compassion is about treating yourself with the same kindness, care, and understanding that you would offer to a dear friend who is suffering. It's about recognising that you are worthy of love and acceptance, even when you make mistakes or experience difficulties.

To cultivate profound self-compassion, begin working with the Inner Critic and Negative Self-Talk, become aware of your inner critic's voice. Notice the words and phrases it uses.
Challenge the validity of your inner critic's statements. Ask yourself if they are truly accurate or helpful. Examine and understand any held beliefs which may be acting as the foundation of a statement that isn't truly serving you. Replace negative self-talk with positive affirmations and self-compassionate statements.

Releasing Old Patterns

The journey inward is not just about acquiring new tools, it is also about shedding what no longer serves us. Old patterns, limiting beliefs, and negative emotions can act as anchors, keeping us tethered to the old Earth paradigm. Releasing these patterns is essential for aligning with the higher frequencies of the new Earth.

Identifying Limiting Beliefs:

Become acutely aware of your thought patterns: Start by simply observing your thoughts without judgment. Are they generally positive or negative? Do they empower you or restrict you? Pay close attention to the self-talk that runs in the background of your mind.
Notice the thoughts that arise in response to specific situations or people.

Challenge the foundation of your assumptions:
Ask yourself: "What evidence do I have to support this belief?"
Consider alternative perspectives. Is there another way to interpret the situation?
Where did this belief originate? Was it instilled in you by your family, culture, or past experiences?
Example: if you think "I am not good enough" ask yourself "compared to what standard? Who set this standard?"

Decipher the "shoulds" and "musts" that constrain you:
These words often reflect rigid rules and expectations that you have imposed on yourself.
Replace "should" with "could" or "choose to."
Example: instead of "I should be more productive," try "I could choose to be more productive, and I also choose to be kind to myself."

Uncover the fear-based beliefs that hold you hostage:
 Fear of failure, fear of rejection, fear of the unknown –
these are common culprits.
Acknowledge the fear, but don't let it dictate your actions.
 Example: if fear of failure stops you from trying something
new, visualise the worst case scenario, then visualise how
you would overcome it.

Journaling as a tool for self-discovery:
 Write about your beliefs, fears, and doubts.
 Explore the connections between your beliefs and your
experiences.
 Use journaling prompts to delve deeper into specific
areas.

Releasing Negative Emotions:

Embrace the full spectrum of your emotions:
>Do not try to suppress or avoid negative emotions.
>They are a natural part of being human.
>Allow yourself to feel the emotion fully, without judgment.
>Recognise that emotions are temporary; they will pass.

Trace the origins of your emotional responses:
>What triggers your anger, sadness, or fear?
>What stories are you telling yourself about these triggers?
>Are these stories based on facts or interpretations?

Example: If a person angers you, is it their action, or your interpretation of that action that is the root of the anger?

Master some powerful emotional release techniques:

Breath work: Practice deep, diaphragmatic breathing to release tension and emotional blockages.

Visualisation: Visualise negative emotions as dark clouds dissipating or as energy flowing out of your body.

EFT (Emotional Freedom Technique): Learn the tapping sequence and use it to release emotional charges associated with specific events or beliefs.

Somatic Experiencing: Allow the body to release stored trauma through gentle movement and awareness.

Nurture yourself with unwavering self-compassion:
Treat yourself with the same kindness and understanding that you would offer to a loved one.
Acknowledge that you are doing your best.
Practice self-soothing techniques such as gentle touch, warm baths, or comforting music.

Breaking Outdated Habits:

Conduct a thorough inventory of your habits:
> Identify the patterns of behaviour that are no
> longer serving you.
> This could include habits related to diet, exercise,
> sleep, relationships, or work.
>
> Example: Do you procrastinate? Do you overspend? Do
> you engage in negative self-talk?

Uncover the underlying triggers that fuel your habits:
> What situations, emotions, or thoughts lead you to engage
> in these habits?
> Are you using these habits to avoid uncomfortable
> feelings or situations?
>
> Example: Do you overeat when you're stressed? Do you
> scroll social media to avoid feeling lonely?

Strategically replace old habits with empowering alternatives:
> Don't just try to eliminate old habits; replace them with
> new, positive ones.
> Choose habits that align with your values and goals.
>
> Example: instead of scrolling social media, try reading a
> book, meditating, or connecting with a friend.

Cultivate unwavering self-discipline:
> Develop the willpower to stick to your new habits, even
> when it's challenging.
> Start small and gradually increase the difficulty.
> Celebrate your successes along the way.

Create a supportive environment for lasting change:
> Surround yourself with people who support your growth.
> Remove temptations and distractions from your
> environment.
> Seek out resources such as books, courses, or coaching.

Shifting from Scarcity to Abundance Mindset:

Dismantle the architecture of scarcity thinking:
> Recognise the pervasive thoughts and beliefs rooted in fear of lack.
> Challenge the assumptions that there is not enough to go around.
> Example: Notice thoughts like "I can't afford that" or "I'll never have enough."

Cultivate a daily practice of gratitude:
> Focus on the abundance that already exists in your life.
> Keep a gratitude journal and write down three things you are grateful for each day.
> Express gratitude to others for their contributions.

Embrace the transformative power of generosity:
> Give freely of your time, talents, and resources.
> Trust that the universe will reciprocate your generosity.
> Example: volunteer your time, donate to a cause you care about, or simply offer a kind word to a stranger.

Visualise a life overflowing with abundance:
> Imagine yourself living a life of abundance in all areas: financial, emotional, relational, and spiritual.
> Use your senses to create a vivid picture of your abundant life.

Embed affirmations of abundance into your subconscious:
> Repeat affirmations such as "I am worthy of abundance," "I am open to receiving abundance," and "Abundance flows to me easily and effortlessly."

Releasing the Need for Control:

Recognise the subtle ways you attempt to control:
> Notice when you try to micromanage situations or people.
> Are you a perfectionist? Do you have difficulty delegating?
> Do you try to predict and control every outcome?

Embrace the liberating power of surrender:
> Trust in the flow of life and let go of the need to control everything.
> Recognise that some things are beyond your control.
> Practice acceptance and allow things to unfold naturally.

Navigate the uncertainty of life with grace:
> Recognise that uncertainty is a natural part of life.
> Develop resilience and adaptability to cope with unexpected changes.
> Focus on what you can control, and let go of what you can't.

Cultivate unwavering trust in yourself and the universe:
> Trust your intuition and inner guidance.
> Trust that the universe is working for your highest good.
> Let go of fear and embrace faith.

Anchor yourself in the present moment through mindfulness:
> Practice mindfulness to stay grounded in the present moment.
> Let go of worries about the future and regrets about the past.
> Focus on what you can control in the present moment.

Connecting with Inner Guidance

Connecting with inner guidance is a multi-faceted journey, requiring consistent practice and a willingness to explore the depths of your consciousness. It's about building a bridge between your conscious mind and the vast reservoir of wisdom within.

Developing Intuition:

Discerning Subtle Intuitive Signals:
Intuition often manifests as a "knowing" that precedes logical thought. It's a sense of certainty prior to forming tangible proof.
Pay attention to micro-expressions and subtle shifts in body language during interactions.
These can provide intuitive insights into others' true feelings.

Become aware of the "clair" abilities:
Clairvoyance (seeing), clairaudience (hearing), clairsentience (feeling), and claircognizance (knowing).
Identify which clairs are strongest for you.
Practice "remote sensing," which involves intuitively perceiving information about a distant location or object.

Refining Intuitive Accuracy:

Keep a detailed "intuition log." Record your intuitive hunches, the circumstances surrounding them, and the outcomes. This helps you identify patterns and refine your accuracy.

Engage in "intuitive games" to test your abilities. Try guessing the colour of a card or the next word someone will say.

Practice "automatic drawing." Allow your hand to move freely across the paper, without conscious control, and interpret the resulting images.

Develop a strong sense of "discernment." Learn to differentiate between genuine intuitive insights and ego-driven desires or fears.

Integrating Intuition into Daily Life:
> Use your intuition to guide your decision-making, both big and small.
> Trust your gut feelings in social interactions and business dealings.
> Practice "intuitive eating," paying attention to your body's hunger cues and cravings.
> Use your intuition to navigate challenging situations and find creative solutions.

Listening to the Heart (Heart-Centred Mastery):

Unveiling the Heart's Deeper Language:
  The heart's wisdom is often expressed through feelings of joy, peace, love, and compassion.
  Learn to recognise the "heart's resistance," which manifests as feelings of anxiety, fear, or unease.
  Practice "heart coherence," aligning your heart rate variability with positive emotions.
  Cultivate a deep sense of "self-love," recognising your inherent worthiness and lovability.

Amplifying Heart-Centred Awareness:
  Practice "heart-focused breathing," visualising your breath flowing in and out of your heart centre.
  Engage in "heart-centred journaling," writing from a place of love and compassion.
  Practice "forgiveness meditation," releasing resentment and anger from your heart.
  Cultivate "heartfelt gratitude," expressing appreciation for the blessings in your life.

Embodying Heart-Centred Living:
  Make choices that are aligned with your heart's values, even when they are unpopular.
  Cultivate authentic relationships based on vulnerability, empathy, and compassion.
  Express your creativity and passions as a way of honouring your heart's desires.
  Practice random acts of kindness.

Aligning with Higher Self (Soul Integration):

Exploring the Higher Self's Dimensions:
The higher self is often described as the "soul's blueprint" or the "divine spark" within. It represents your highest potential and your soul's purpose.
Connecting with your higher self can bring a sense of deep peace, purpose, and fulfilment.

Establishing a Profound Connection:
Practice "soul retrieval," reclaiming lost aspects of your soul.
Engage in "past life regression," exploring previous incarnations to gain insights into your soul's journey.
Practice "Akashic Records readings," accessing the records of your soul's experiences.
Cultivate a strong sense of "spiritual connection," recognising your interconnectedness with all of life.

Manifesting Higher Self Alignment:
Live your life with integrity, authenticity, and purpose.
Use your unique gifts and talents to serve others and make a positive impact on the world.
Embrace your "soul's mission," aligning your life with your highest purpose.
Practice daily acts of service that are aligned with your souls purpose.

Advanced Techniques for Inner Guidance:

Channeling:
> Learn to become a clear channel for higher guidance, allowing information to flow through you.
> Practice "automatic speech," allowing your voice to express messages from your higher self or spirit guides.
> Develop the ability to channel specific entities or energies.

Energy Healing:
> Learn to work with energy to clear blockages and enhance connection with your inner guidance.
> Practice Reiki, Qi Gong, or other energy healing modalities.
> Develop the ability to sense and manipulate energy fields.

Dream Work:
> Become very proficient at lucid dreaming.
> Learn to decode and interpret complex dream symbolism.
> Use dreams as a portal for communication with your higher self.

By dedicating yourself to these practices, you can cultivate a profound and unwavering connection with your inner guidance, empowering you to navigate life's challenges with clarity, confidence, and grace.

Self-Mastery

Self-mastery is the ongoing journey of aligning with your highest potential, integrating inner tools, and living as a sovereign being. It's about becoming a conscious creator of your reality.

Integrating All Tools and Practices :

Creating a Personalised Inner Alchemy:
Develop a daily ritual that seamlessly blends meditation, mindfulness, affirmations, and gratitude practices.
Utilise journaling to track patterns, insights, and emotional shifts, refining your approach over time.
Incorporate physical practices like yoga, tai chi, or breath work to anchor inner work in the body.

Dynamic Adaptation:
Recognise that your needs evolve. Regularly reassess your practices and adjust them accordingly.
Embrace periods of "inner silence" to allow for intuitive recalibration of your routine.
View challenges as opportunities to apply and refine your integrated tools.

The Power of Habit and Intention:
Establish firm intentions for your inner work, reinforcing your commitment.
Leverage the power of habit to create consistency, making inner work a natural part of your day.
Use reminders, alarms, or visual cues to support your daily practice.

Embodying Higher Consciousness (Advanced Practice):

Cultivating Heart-Brain Coherence:
> Practice techniques to synchronise heart rhythms with brainwave states, fostering inner harmony.
> Develop the ability to access and maintain states of flow, where intuition and creativity flourish.
> Use heart-centred visualisation to imbue your thoughts and actions with love and compassion.

Conscious Frequency Modulation:
> Become acutely aware of your vibrational state and learn to shift it intentionally.
> Use techniques like chanting, toning, or visualisation to raise your frequency and align with higher timelines.
> Practice transmuting negative emotions into positive ones through conscious awareness.

Ego Transcending Mastery:
> Develop the ability to observe egoic patterns without judgment, allowing them to dissolve naturally.
> Cultivate a deep sense of detachment from outcomes, trusting in the unfolding of divine timing.
> Practice radical self-acceptance, embracing all aspects of yourself with compassion.

Living as a Sovereign Being (Empowered Presence):

Energetic Sovereignty:
 Develop the ability to sense and manage your energy field, becoming impervious to external negativity.
 Practice techniques for grounding, shielding, and clearing your energy.
 Learn to discern between your own energy and the energy of others.

Boundary Setting and Assertiveness:
 Clearly define your boundaries and communicate them with confidence and clarity.
 Practice assertive communication, expressing your needs and desires without aggression or passivity.
 Develop the ability to say "no" without guilt or explanation.

Inner Authority and Self-Guidance:
 Cultivate a deep trust in your intuition and inner guidance, becoming your own source of wisdom.
 Develop the ability to discern between genuine intuition and ego-driven impulses.
 Practice self-inquiry, exploring your beliefs and values to align with your authentic truth.

Conscious Creation:
 Learn the laws of attraction, and how to use them to manifest your desired reality.
 Practice visualisation, and intention setting.
 Recognise, and remove, subconscious blocks that are preventing manifestation.

Advanced Practices for Self-Mastery:

Conscious Dreaming and Reality Shifting:
   Explore the nature of reality through lucid dreaming,
   gaining insights into the subconscious and
   multidimensional realms.
   Practice reality shifting techniques to consciously alter
   your experience of reality.
   Use dreams as a portal for healing, manifestation, and
   spiritual exploration.

Energy Mastery and Subtle Body Awareness:
   Develop a deep understanding of the subtle energy
   bodies and their functions.
   Practice techniques for clearing, balancing, and activating
   the chakras and meridians.
   Explore the use of crystals, essential oils, and sound
   healing to enhance energy flow.

Pranic Nourishment and Light Body Activation:
   Explore the possibility of living on prana (life force
   energy), gradually reducing reliance on physical food.
   Practice techniques for activating the light body,
   enhancing your connection to higher dimensions.
   This is a very advanced practice that demands extreme
   care, and dedication.

Sacred Geometry and Cosmic Alignment:
   Study the principles of sacred geometry to understand
   the underlying patterns of creation.
   Use sacred geometry to create sacred spaces, enhance
   energy flow, and align with cosmic frequencies.
   Explore the connection between sacred geometry and the
   human energy field.

The Fruits of Self-Mastery (Transcendental Living):

Unconditional Peace and Joy:
> Experience a deep sense of inner peace and joy that is independent of external circumstances.
> Cultivate a state of unwavering equanimity, remaining centred amidst life's challenges.

Harmonious Co-Creation:
> Contribute to the creation of a more conscious and harmonious world through your presence and actions.
> Become a beacon of light, inspiring others to awaken to their own potential.

Multidimensional Awareness:
> Develop the ability to perceive and interact with multiple dimensions of reality.
> Access higher realms of consciousness and receive guidance from spiritual beings.

Timeless Presence:
> Live fully in the present moment, free from the constraints of past regrets or future anxieties.
> Experience a sense of timelessness and unity with all of life.

By embracing these advanced practices and cultivating unwavering dedication, you can embark on a journey of profound self-mastery, transforming your life and contributing to the evolution of consciousness.

Conscious Communication

Conscious communication is the art of expressing ourselves with intention, listening with empathy, and fostering understanding in our interactions. It's about moving beyond superficial exchanges and creating authentic connections that honour the dignity of all beings.

Speaking with Intention:

Clarity and Authenticity:
>	Speak your truth with clarity and honesty.
>	Express your thoughts and feelings in a way that is authentic to who you are.
>	Avoid using vague or misleading language.

Mindful Word Choice:
>	Be mindful of the power of your words.
>	Choose words that uplift, inspire, and empower.
>	Avoid using language that is critical, judgmental, or harmful.

Purposeful Expression:
>	Consider the purpose of your communication.
>	What message do you want to convey?
>	What outcome do you desire?
>	Speak with a clear intention in mind.

Taking Responsibility:
>	Understand that your words have an impact.
>	Take responsibility for the energy you bring to your communication.
>	If you cause harm, be prepared to make amends.

Active Listening:

Presence and Attention:
>Give your full attention to the person who is speaking.
>Be present in the moment and avoid distractions.
>Make eye contact and use nonverbal cues to show that you are listening.

Empathy and Understanding:
>Listen with empathy and try to understand the other person's perspective.
>Put yourself in their shoes and see the world through their eyes.
>Suspend judgment and avoid interrupting.

Reflective Listening:
>Reflect back what you have heard to ensure understanding.
>Paraphrase the other person's words and summarise their main points.
>Ask clarifying questions to deepen your understanding.

Listening to Nonverbal Cues:
>Pay attention to the other person's body language, tone of voice, and facial expressions.
>Nonverbal cues can often reveal more than words alone.
>Allow silence, do not feel like you must fill it.

Fostering Understanding:

Open and Honest Dialogue:
> Create a safe space for open and honest dialogue.
> Encourage the other person to share their thoughts and feelings.
> Be willing to share your own vulnerabilities.

Respectful Communication:
> Communicate with respect, even when you disagree.
> Avoid personal attacks and name-calling.
> Focus on the issues at hand, rather than the person.

Seeking Common Ground:
> Look for common ground and areas of agreement.
> Focus on solutions and collaboration, rather than conflict.
> Be willing to compromise and find win-win solutions.

Bridging Differences:
> Acknowledge and respect differences in perspectives and experiences.
> Seek to understand the root causes of disagreements.
> Build bridges of understanding through empathy and compassion.

Using "I" Statements
> When needing to express a difficult subject, use "I" statements.
> Example: instead of "You make me mad when..." use "I feel frustrated when..."

Mindful Communication Exercises:
> Practice speaking and listening mindfully in everyday conversations.
> Set intentions for conscious communication before engaging in conversations.
> Reflect on your communication patterns and identify areas for improvement.

Nonviolent Communication (NVC):
> Learn the principles of NVC, which emphasises empathy, honesty, and compassion.
> Use NVC to resolve conflicts and build stronger relationships.

Mediation and Conflict Resolution:
> Develop skills in mediation and conflict resolution.
> Learn how to facilitate productive conversations and resolve disagreements peacefully.

Journaling and Self-Reflection:
> Use journaling to explore your communication patterns and identify triggers.
> Reflect on your interactions and identify areas where you can improve.

Conscious communication is a powerful tool for building stronger relationships, resolving conflicts, and creating a more harmonious world. By speaking with intention, listening with empathy, and fostering understanding, we can create a culture of communication that honours the dignity of all beings.

Building Conscious Relationships

Conscious relationships are intentional partnerships that go beyond societal norms, fostering deep soul connection, personal growth, and co-creation. They are laboratories for love, where authenticity, respect, and mutual support are the cornerstones.

Cultivating Profound Authenticity (Radical Transparency):

Shadow Work and Integration:
Embrace your shadow self, acknowledging and integrating your darker aspects.
Share your vulnerabilities and fears, creating a space for mutual healing.
Practice radical honesty, even when it's uncomfortable.

Energetic Alignment:
Become aware of your energetic presence and how it impacts your partner(s).
Cultivate energetic transparency, allowing your partner(s) to sense your true intentions.
Practice energetic clearing to release blockages and maintain authentic connection.

Authentic Expression of Needs:
Clearly articulate your needs, desires, and boundaries without fear of judgment.
Encourage your partner(s) to do the same, creating a space for mutual understanding.
Practice non-violent communication techniques to express needs effectively.

Fostering Deep Respect (Sacred Partnership):

Honouring Soul Sovereignty:
> Recognise and respect your partner(s) as sovereign beings with their own unique path.
> Avoid projecting your expectations or desires onto them.
> Support their individual growth and evolution, even if it diverges from your own.

Creating Sacred Space:
> Designate time and space for conscious connection, free from distractions.
> Practice rituals and ceremonies that honour the sacredness of your partnership.
> Cultivate a sense of reverence for each other's presence.

Active Empathic Listening:
> Go beyond surface-level listening and truly hear your partner's emotional landscape.
> Practice empathic mirroring, reflecting back their feelings and experiences.
> Validate their emotions without trying to fix or change them.

Nurturing Profound Mutual Support (Co-Evolution):

Shared Purpose and Vision:
> Co-create a shared vision for your relationship, aligning on your values and goals.
> Engage in collaborative projects and activities that support your shared purpose.
> Become a team that works together to serve a higher purpose.

Emotional Resonance and Co-Regulation:
> Develop the ability to sense and respond to each other's emotional states.
> Practice co-regulation techniques, helping each other to manage stress and anxiety.
> Create a safe haven for emotional vulnerability and healing.

Spiritual Partnership:
> Engage in shared spiritual practices, such as meditation, prayer, or energy work.
> Support each other's spiritual growth and exploration.
> Recognise the divine presence within each other.

Conscious Conflict Resolution:
> View conflict as an opportunity for growth.
> Practice deep listening, and seek to understand the underlying needs of all parties involved.
> Use conflict as a catalyst for deeper intimacy.

Advanced Tools and Practices (Transformative Connection):

Tantric Practices:
>Explore tantric techniques to deepen intimacy and cultivate energetic connection.
>Practice conscious sexuality as a path to spiritual growth and union.
>Learn to harness sexual energy for healing and transformation.

Energy Exchange and Clearing:
>Become aware of the energetic exchange that occurs in relationships.
>Practice techniques for clearing and balancing each other's energy fields.
>Use energy healing to release emotional blockages and enhance connection.

Soul Retrieval and Integration:
>Support each other in the process of soul retrieval, reclaiming lost aspects of the self.
>Practice integration techniques to harmonise fragmented parts of the psyche.
>Create a space for mutual soul growth and wholeness.

Multidimensional Relationship Awareness:
>Recognise that relationships exist on multiple dimensions.
>Become aware of past life connections, and karmic patterns.
>Work together to clear any negative karmic bonds.

By embracing these advanced practices and cultivating a deep commitment to conscious connection, you can create relationships that are truly transformative, serving as a catalyst for personal and collective evolution.

Ethical Action and Service

Ethical action and service, at its core, is the embodiment of conscious evolution. It's the transition from passive observer to active co-creator, where every thought, word, and deed is infused with intention and aligned with the highest good.

Aligning Actions with Values:

Cellular Ethics:
> Extend your ethical awareness to the very cellular level. Consider the impact of food choices, environmental toxins, and even thought patterns on your physical and energetic well-being.
> Practice "cellular gratitude," acknowledging the intricate intelligence of your body and its interconnectedness with the Earth.

Relational Ethics:
> Cultivate radical transparency in all relationships, fostering a culture of authenticity and trust.
> Practice "conscious conflict resolution," viewing disagreements as opportunities for deeper understanding and growth.
> Develop the ability to discern and honour the energetic boundaries of others.

Environmental Ethics:
> Embrace a deep sense of stewardship for the Earth, recognising its intrinsic value and interconnectedness with all life.
> Practice "regenerative living," actively contributing to the restoration and revitalisation of ecosystems.
> Advocate for policies that protect biodiversity, reduce pollution, and promote sustainable resource management.

Cosmic Ethics:

Extend your ethical considerations beyond the Earth, recognising our interconnectedness with the cosmos. Practice "planetary citizenship," acknowledging our responsibility to contribute to the well-being of the entire solar system and beyond.

Cultivate a sense of awe and wonder for the vastness and complexity of the universe.

Contributing to the Greater Good (Systemic Transformation):

Conscious Economics

Advocate for economic systems that prioritise social and environmental well-being over profit maximisation.
Support businesses that are committed to fair labor practices, sustainable resource management, and community development.
Explore alternative economic models, such as the circular economy and the gift economy.

Conscious Governance

Embodying Higher Consciousness Leadership:
Prioritise the emergence of leaders who have demonstrably integrated higher vibrational templates, exhibiting qualities of wisdom, compassion, and intuitive clarity.
Support systems that recognise and elevate individuals who operate from a foundation of unity consciousness, prioritising the well-being of the whole.

Facilitating Conscious Decision-Making:
Advocate for governance models that incorporate intuitive guidance, collective consciousness input, and alignment with universal principles.
Support the implementation of technologies and practices that amplify conscious awareness in decision-making processes.

Guiding the Collective Shift:
Champion leaders who understand the dynamics of the ascension process and are committed to guiding humanity through the shift with grace and wisdom.
Support initiatives that educate and empower individuals to participate in conscious co-creation, fostering a collective understanding of the new Earth paradigm.

Conscious Media

Support media outlets that are committed to truth, accuracy, and ethical reporting.
Cultivate media literacy, discerning between credible sources and misinformation.
Use social media platforms to promote positive messages and inspire conscious action.

Conscious Education

Cultivating Higher Consciousness Embodiment:
Prioritise educational models that nurture the development of individuals who embody higher vibrational templates, emphasising intuitive wisdom,
compassionate understanding, and conscious awareness.
Support educators who guide students in integrating spiritual intelligence and aligning with universal principles.

Fostering Conscious Co-Creation:
Advocate for educational systems that empower individuals to understand and participate in the conscious co-creation of the New Earth paradigm.
Support the integration of technologies and practices that amplify conscious awareness, intuitive abilities, and anchoring collective wisdom.

Guiding Evolutionary Advancement:
Champion educational leaders who comprehend the ascension process and are dedicated to guiding humanity's evolutionary advancement.
Support initiatives that promote lifelong conscious evolution, self-discovery, and the integration of higher dimensional perspectives.

Serving Others (Radical Compassion - Consciousness Alignment):

Intersectional Compassion (Unity Consciousness Realization):
Cultivate a deep realisation of unity consciousness, recognising the interconnectedness of all beings and the illusory nature of separation.
Shift internal beliefs that perpetuate marginalisation, focusing on dissolving the thought patterns that create division.
Embody the vibration of inclusion, radiating acceptance and understanding to dissolve the energetic roots of oppression.

Trauma-Informed Service (Healing Consciousness Imprints):
Recognise trauma as a consciousness imprint, understanding its energetic impact on individuals and collectives.
Practice holding space for healing, radiating frequencies of love and compassion to transmute trauma's energetic residues.
Advocate for the transformation of collective consciousness, shifting the belief systems that perpetuate trauma.

Ecological Restoration Service (Gaia Consciousness Harmony):
Align with Gaia's consciousness, recognising the planet as a sentient being and our interconnectedness with all life.
Cultivate beliefs of harmony and balance, visualising and intending the restoration of ecosystems through conscious alignment.
Embody the frequency of reverence for nature, fostering a deep energetic connection with the Earth.

Spiritual Activism (Consciousness Transformation Catalysis):
Integrate spiritual practices as tools for consciousness transformation, cultivating inner peace and radiating higher frequencies.
Use meditation, visualisation, and intention to shift collective consciousness, aligning with higher timelines and possibilities.

Recognise activism as a form of inner work, transforming beliefs and radiating frequencies that shape the holographic reality.

Advanced Tools and Practices (Ethical Alchemy):

Shadow Integration for Collective Healing:
Recognise that collective shadow aspects (prejudice, greed, violence) must be brought to light and healed.
Practice personal shadow work in order to better understand and transmute these collective shadows.

Time-Line Shifting Ethical Choices:
Understand that every choice creates a ripple through time, and effects future time-lines.
Practice making choices that create the most positive time-line for all beings.

Ethical Code Creation:
Create a personal ethical code that is a living document, and that is regularly updated.
Create ethical codes for your business, or for organisations you are part of.

Conscious Ancestral Healing:
Recognise that ancestral patterns and traumas can influence our ethical choices.
Engage in ancestral healing practices to release negative patterns and cultivate a legacy of ethical living.

By embracing this perspective, you can become a true ethical alchemist, transforming yourself and the world around you through conscious action and service.

Co-Creating a New Earth

Co-creating a New Earth at this level transcends conventional understanding, delving into the realm of quantum consciousness, inter-dimensional collaboration, and the conscious evolution of planetary systems. It's about becoming active participants in the unfolding cosmic drama, weaving a tapestry of light, love, and infinite potential.

Collective Action and Collaboration (Fractal Connectivity):

Morphogenetic Field Resonance:
> Understand and utilise morphogenetic fields, the invisible templates that shape reality, to amplify collective intentions.
> Practice resonant field techniques, aligning individual consciousness with the planetary morphogenetic field.
> Create collective fields of intention for healing, peace, and abundance.

Symbiotic Tech Ecosystems:
> Develop technologies that seamlessly integrate with natural ecosystems, enhancing biodiversity and ecological balance.
> Create symbiotic Intelligence systems that collaborate with human consciousness to solve complex global challenges.
> Implement holographic communication networks that facilitate instantaneous and intuitive connection.

Time-Traveling Collaboratives:
> Explore the potential of time-travel technologies for learning from past civilisations and influencing future timelines.
> Create time-travel collaboratives that work to heal historical traumas and prevent future catastrophes as we continue to evolve.
> Develop ethical guidelines for time travel, ensuring responsible use of this powerful technology.

Extra-Planetary Alliances:

Establish communication and collaboration with extra-planetary civilisations.

Engage in intergalactic councils to address cosmic challenges and share universal wisdom.

Participate in planetary ascension projects with extra-planetary support.

<u>Shared Vision and Purpose (Quantum Narratives):</u>

Holographic Reality Projections:
  Utilise holographic technologies to create immersive experiences of a New Earth, inspiring collective vision and action.
  Develop holographic reality projections that allow individuals to experience different future timelines.
  Create interactive holographic environments that facilitate collaborative design and problem-solving.

Consciousness-Based Architecture:
  Design buildings and cities that resonate with specific frequencies, promoting healing, creativity, and spiritual growth.
  Incorporate bio-luminescent materials and living architecture that adapts to the needs of its inhabitants.
  Create sacred spaces that amplify collective intentions and facilitate inter-dimensional communication.

Quantum Language and Symbolism:
  Develop a quantum language that transcends linear communication, facilitating instantaneous understanding and telepathic connection.
  Utilise quantum symbols that encode complex information and activate higher consciousness.
  Create quantum art that transmits healing frequencies and inspires collective awakening.

Multi-Sensory Storytelling:
  Develop multi-sensory storytelling experiences that engage all aspects of human consciousness.
  Incorporate virtual reality, augmented reality, and haptic technologies to create immersive narratives.
  Use soundscapes, light shows, and aroma therapy to enhance the emotional impact of stories.

Building a Harmonious Future (Hyper-Dimensional Systems):

Zero-Point Energy Systems:
> Develop technologies that harness zero-point energy, providing clean and abundant energy for all.
> Implement zero-point energy grids that are decentralised and resilient.
> Create zero-point energy healing devices that restore cellular health and vitality.

Planetary Consciousness Networks:
> Develop technologies that connect individual consciousness to the planetary consciousness network.
> Create planetary consciousness hubs that facilitate collective meditation and intention-setting.
> Implement planetary consciousness monitoring systems that track global emotional and energetic states.

Inter-Dimensional Healing Chambers:
> Develop inter-dimensional healing chambers that utilise advanced energy healing technologies.
> Create healing chambers that facilitate cellular regeneration, DNA activation, and consciousness expansion.
> Implement inter-dimensional healing protocols that address the root causes of disease and suffering.

Time-Line Re-Calibration Technologies:
> Develop technologies that allow for the re-calibration of planetary timelines, shifting towards more positive outcomes.
> Create time-line re-calibration protocols that are guided by ethical principles and collective wisdom.
> Implement time-line re-calibration projects that address historical traumas and prevent future catastrophes.

Advanced Tools and Practices (Cosmic Alchemy):

DNA Activation and Light Body Integration:
> Practice advanced DNA activation techniques to unlock dormant potentials and enhance consciousness.
> Integrate the light body, the energetic blueprint of the soul, into the physical body.
> Develop technologies that facilitate DNA activation and light body integration.

Consciousness Transference and Embodiment:
> Explore the potential of consciousness transference, allowing for the seamless transfer of consciousness between different bodies.
> Develop technologies that facilitate consciousness embodiment in virtual realities and other dimensions.
> Create ethical guidelines for consciousness transference, ensuring responsible use of this technology.

Inter-Galactic Seedship Technologies:
> Develop inter-galactic seedship technologies that facilitate the seeding of life on other planets.
> Create seedship protocols that ensure the responsible and ethical introduction of life to other worlds.
> Implement seedship projects that contribute to the expansion of life throughout the cosmos.

Cosmic Resonance Healing:
> Practice techniques that align human resonance with cosmic resonance.
> Use sound, light, and vibration to bring the body into harmonic resonance with the cosmos.
> Create cosmic resonance healing centres that integrate advanced technologies and ancient wisdom.

At this level, "Co-Creating a New Earth" becomes a journey of cosmic alchemy, where we transcend the limitations of the physical world and embrace our role as conscious co-creators of the universe.

Embodiment of the New Paradigm
(Hyper-Dimensional Integration)

 Embodiment of the New Paradigm is the art of weaving higher consciousness into the fabric of our existence, becoming living portals for the manifestation of a new reality. It's about transcending the limitations of duality and anchoring the multidimensional self into the physical realm, radiating a transformative influence into the collective consciousness.

Living as a Beacon of Light (Quantum Radiance):

Aura Expansion and Luminosity:
> Practice techniques to expand and purify your aura, becoming a radiant field of light.
> Cultivate the ability to consciously modulate your auric field, projecting specific frequencies for healing and inspiration.
> Develop the capacity to perceive and interact with the auric fields of others, offering energetic support and guidance.

Heart-Brain Coherence Resonance:
> Master the art of heart-brain coherence, synchronising heart rhythms and brainwave states for optimal functioning.
> Cultivate a state of sustained heart-centred awareness, radiating compassion and wisdom.
> Develop the ability to transmit coherent heart frequencies, creating a ripple effect of peace and harmony.

Light Language Embodiment:

Activate and embody light language, the universal language of consciousness, through sound, gesture, and intention.

Use light language to transmit encoded information, activate dormant DNA, and facilitate consciousness expansion.

Develop the ability to translate light language into verbal communication, conveying complex spiritual concepts.

Multi-Dimensional Presence:

Cultivate awareness of your multidimensional self, integrating your higher aspects into your physical embodiment.

Practice techniques for grounding and anchoring your higher self, radiating its qualities into your daily life.

Develop the ability to shift your consciousness between different dimensions, accessing higher realms of wisdom and inspiration.

Inspiring Others to Awaken (Catalytic Influence):

Consciousness Mentoring:
> Develop the ability to guide and support others on their path of awakening, acting as a catalyst for their transformation.
> Create personalised consciousness mentoring programs, tailoring guidance to the unique needs of each individual.
> Utilise intuitive guidance and energetic Organise to facilitate profound shifts in consciousness.

Sacred Activism:
> Integrate spiritual principles into activism, creating movements that are grounded in love, compassion, and nonviolence.
> Organise sacred activism events, such as peace meditations, healing ceremonies, and environmental restoration projects.
> Develop strategies for conscious activism that address the root causes of social and environmental challenges.

Consciousness-Based Media:
> Create media content that raises consciousness, inspires positive change, and promotes the values of the New Paradigm.
> Develop interactive media platforms that facilitate collective dialogue and co-creation.
> Utilise immersive technologies to create experiences that awaken empathy and inspire action.

Planetary Grid Work:
> Participate in planetary grid work, anchoring light and positive intentions into the Earth's energy field.
> Develop the ability to perceive and interact with planetary grids, identifying areas that require healing and support.
> Organise global grid work meditations and ceremonies, uniting consciousness for planetary transformation.

Anchoring Higher Frequencies (Transcendental Grounding):

Crystalline Body Activation:
> Activate the crystalline structure within your body, enhancing your ability to hold and transmit higher frequencies.
> Practice techniques for clearing and aligning your chakras, meridians, and energy bodies.
> Develop the ability to consciously modulate the crystalline matrix of your body.

Gaia Consciousness Resonance:
> Cultivate a deep connection with Gaia, the living consciousness of the Earth, and resonate with her frequencies.
> Practice Gaia meditation, attuning your consciousness to the Earth's heartbeat and rhythms.
> Develop the ability to communicate with Gaia, receiving guidance and support for planetary stewardship.

Inter-Dimensional Portals:
> Create inter-dimensional portals in your home and community, anchoring higher frequencies and facilitating inter-dimensional communication.
> Use sacred geometry, crystals, and sound to create portals that resonate with specific dimensions.
> Develop the ability to consciously travel through inter-dimensional portals, accessing higher realms of consciousness.

Advanced Tools and Practices (Cosmic Integration):

Time-Line Weaving:
>   Develop the ability to consciously weave timelines, creating positive futures for yourself and the planet. Practice time-line visualisation and intention-setting, aligning your consciousness with desired outcomes. Develop the ability to perceive and interact with multiple timelines, choosing the most harmonious path.

Cosmic Resonance Healing:
>   Practice techniques that align your resonance with cosmic frequencies, facilitating deep healing and transformation. Use sound, light, and vibration to bring your body into harmonic resonance with the cosmos. Develop the ability to transmit cosmic resonance healing to others, facilitating cellular regeneration and DNA activation.

Inter-Galactic Ambassador:
>   Develop the ability to communicate and collaborate with inter-galactic civilisations, acting as an ambassador for humanity. Practice inter-galactic communication protocols, ensuring clear and respectful dialogue. Develop the ability to share Earth's wisdom and receive guidance from inter-galactic beings.

Planetary Ascension Mastery:
>   Master the art of planetary ascension, guiding and supporting the Earth's transition to a higher dimensional reality. Practice advanced ascension techniques, such as light body activation, DNA recalibration, and planetary grid work. Develop the ability to perceive and interact with the Earth's ascension process, offering energetic support and guidance.

By embodying these hyper-dimensional practices, you become a living embodiment of the New Paradigm, radiating a transformative influence that ripples through the cosmos.

Sustainable Living Systems (Hyper-Dimensional Ecosystem Integration)

Hyper-dimensional ecosystem integration represents a paradigm shift towards a symbiotic co-creation with the Earth, where human activities not only sustain but actively enhance the planet's energetic and ecological integrity. It's about transcending the limitations of 3D models and embracing a multi-dimensional approach that harmonises with the Earth's subtle energies and cosmic rhythms.

Hyper-Regenerative Agriculture (Quantum Food Webs):

Quantum Agriculture and Bio-field Resonance:
>Utilise quantum entanglement and bio-field resonance to enhance plant growth and nutrient uptake.
>Develop quantum agriculture technologies that optimise seed germination and crop yields.
>Implement bio-field resonance techniques to enhance soil fertility and microbial diversity.

Holographic Food Synthesis:
>Explore the potential of holographic food synthesis, creating nutrient-rich food from light and energy.
>Develop holographic food printers that can create personalised meals based on individual's *current* nutritional needs.

Symbiotic Intelligence-Driven Ecosystem Management:
>Utilise Intelligence-driven systems that learn and adapt to the complex dynamics of ecosystems.
>Develop Intelligence-powered orbs that can monitor and manage entire agricultural landscapes.
>Implement Intelligence-Driven predictive models that anticipate and mitigate environmental challenges.

Inter-Species Communication and Co-Creation:
     Establish communication protocols with plants and
     animals, fostering a deeper understanding of their needs.
     Develop technologies that facilitate inter-species dialogue
     and collaboration.
     Implement co-creation projects with other species,
     designing ecosystems that benefit all life.

Inter-dimensional Renewable Energy (Cosmic Energy Harvesting):

Zero-Point Energy Harvesting and Distribution:
     Develop technology/understanding that taps into the zero-
     point energy field, accessing clean and abundant energy.
     Implement zero-point energy grids that are decentralised
     and self-sustaining.
     Utilise zero-point energy devices for personal and
     community use.

Space-Based Energy Storage:
     Implement space-based energy storage systems, utilising
     advanced technologies like antimatter storage.

Galactic Energy Grid Integration:
     Explore the potential of integrating with the galactic
     energy grid, accessing cosmic energy sources.
     Develop technologies that can tap into the energy flows of
     the galaxy.
     Implement galactic energy transmission protocols for
     inter-planetary collaboration.

Consciousness-Driven Energy Modulation:
     Develop technologies that allow individuals to modulate
     energy flow through conscious intention.
     Implement consciousness-driven energy healing devices
     that restore balance and harmony.
     Utilise consciousness-driven energy systems for personal
     and community empowerment.

Hyper-Ecosystem Restoration (Planetary Regeneration):

Planetary Grid Healing and Activation:
   Cultivate methods that can heal and activate the Earth's planetary grids, restoring energetic balance.
   Implement planetary grid healing protocols that address global environmental challenges.
   Utilise planetary grid activation techniques to enhance consciousness and promote harmony.

Inter-Dimensional Ecosystem Restoration:
   Explore the potential of restoring ecosystems in other dimensions and realms.
   Develop techniques that facilitate inter-dimensional ecosystem restoration.
   Implement inter-dimensional ecosystem management protocols.

Atmospheric Bio-Engineering and Terraforming:
   Develop practices for atmospheric bio-engineering, restoring the Earth's atmosphere to its optimal state.
   Explore the potential of terraforming other planets, creating habitable environments for life.
   Implement ethical guidelines for atmospheric bio-engineering and terraforming.

Consciousness-Based Ecosystem Co-Creation:
   Develop technologies that allow individuals to co-create ecosystems through conscious intention.
   Implement consciousness-based ecosystem design protocols.
   Utilise consciousness-based ecosystem restoration techniques.

Hyper-Dimensional Tools and Practices (Cosmic Stewardship):

Holographic Ecosystem Modelling and Simulation:
> Develop holographic ecosystem modelling and simulation technologies, allowing for the creation and testing of complex ecosystems.
> Implement holographic ecosystem design tools for urban planning and infrastructure development.
> Utilise holographic ecosystem simulations for educational purposes.

Quantum Communication and Inter-Species Collaboration:
> Develop quantum communication technologies that facilitate instantaneous communication with all life forms.
> Implement quantum communication protocols for inter-species collaboration.
> Utilise quantum communication for ecosystem monitoring and management.

Consciousness-Driven Weather Modification:
> Explore the potential of consciousness-driven weather modification, harmonising with natural weather patterns.
> Develop systems that can amplify and direct conscious intention for weather modification.
> Implement ethical and social guidelines for consciousness-driven weather modification.

Galactic Ecosystem Stewardship:
> Develop protocols for galactic ecosystem stewardship, ensuring the responsible management of planetary and galactic resources.
> Implement intergalactic collaboration projects for ecosystem restoration and protection.
> Utilise galactic ecosystem monitoring systems for early detection of environmental challenges.

By embracing these hyper-dimensional ecosystem integrations, we can transcend the limitations of conventional sustainability and become true stewards of the Earth and the cosmos.

Glossary for following section

DAO (Decentralised Autonomous Organisation):
> An organisation represented by rules encoded as a program. It's transparent, controlled by members, and not influenced by central authority.

DLT (Distributed Ledger Technology):
> A decentralised database managed by multiple participants. (Blockchain is a type of DLT.)

Holacracy:
> A decentralised organisational structure in which authority and decision-making are distributed throughout self-organising teams ("circles") rather than being vested in a hierarchy.

Sociocracy:
> A governance system that uses circles and consent-based decision-making to create equitable and effective organisations.

Universal Basic Experiences (UBE):
> Access to enriching experiences and personal growth opportunities for all, replacing the concept of monetary-based universal basic income.

Conscious Communities and Governance

(Hyper-Integrated & Contextualised Models - Post-Singularity, Post-Finance, Love-Governed)

In a reality where material needs are met through free creation and governance is aligned with the energetic frequency of love, conscious communities become expressions of collective flourishing and harmonious resonance, operating on principles of unity and abundance.

Collaborative Living (Synergistic & Evolving Community Ecosystems):

Eco-Villages with Sociocratic Structures & Permaculture Integration:
> Communities utilise advanced bio-fabrication technologies to create dwellings and infrastructure from living materials, seamlessly integrating with the natural environment.

> Sociocratic circles engage in "resonance mapping," using biofeedback and holographic projections to visualise the energetic impact of decisions on the community and its ecosystem.

> Data visualisation extends to mapping the flow of subtle energies and consciousness within the community, fostering a deeper understanding of interconnectedness.

Co-Housing with Holacracy & Enhanced Role Optimisation:
> Roles are fluid and adaptable, evolving organically based on individual passions and the collective's evolving needs.

> "Tactical" and "governance" gatherings are transformed into "creative flow sessions," where participants engage in collaborative brainstorming and intuitive problem-solving.

> Holographic interfaces provide real-time feedback on the community's energetic coherence, allowing members to adjust their contributions for optimal harmony.

Intergenerational Communities with Restorative Justice & Holographic Storytelling:

Restorative justice circles utilise "heart-field alignment" techniques, allowing participants to experience the energetic impact of their actions on others.

Holographic storytelling becomes a form of collective memory weaving, preserving a community's history and wisdom in a dynamic and interactive format.

Decentralised Systems (Hyper-Connected & Resilient Networks - Energetic Flow):

Blockchain for Transparent Governance & Distributed Ledger Technology (DLT) for Energetic Flow Tracking:

Blockchain records "energetic agreements," ensuring that all interactions are aligned with the frequency of love and transparency.

DLT tracks the flow of "creative energy," providing insights into the community's collective output and areas for enhanced collaboration.

Decentralised Autonomous Organisations (DAOs) with Resonance-Based Governance:

DAOs operate on "heart-field consensus," where decisions are made by attuning to the collective's energetic resonance.

Dashboards visualise "energetic coherence" and "creative flow," providing real-time feedback on the community's well-being.

Peer-to-Peer Networks with Mesh Networking & Quantum Communication:

Networks facilitate the "instantaneous transmission of heart-centred intentions," transcending the limitations of distance and time.

Quantum communication ensures the secure and reliable exchange of "love-encoded data."

Enhanced Skill Matching & Universal Basic Experiences (UBE):

Skill matching connects individuals with opportunities to express their "unique creative frequencies."

UBE programs provide access to "transformative experiences," fostering personal growth and collective evolution.

Participatory Decision-Making (Collective Intelligence & Conscious Evolution - Heart-Centred Processes):

Citizen Assemblies with Deliberative Resonance & Temporary Reality (TR) Simulations:
   TR Simulations create "empathy immersion experiences," allowing participants to feel the energetic impact of decisions on all stakeholders.

   Decisions are made through "heart-field resonance testing," ensuring alignment with the highest good of the community.

Open Source Governance Platforms with Resonance Analysis & Citizen Feedback Loops:
   Platforms analyse the "energetic signature" of proposed initiatives, ensuring alignment with the community's values.

   Feedback loops measure the "energetic ripple effect" of decisions, allowing for continuous optimisation.

Consensus-Based Decision-Making with Nonviolent Communication & Heart-Field Integration:
   Heart-field integration allows participants to "feel the truth" of others' perspectives, fostering deep understanding and compassion.

   Decisions are made through "collective heart-field attainment," ensuring alignment with the frequency of love.

'Online' Platforms for Collaborative Co-Creation & Augmented Reality (AR) Visualisation:
   AR overlays visualise the "energetic blueprints" of creative projects, enhancing collaboration and shared vision.

   Platforms facilitate "collective dream weaving," allowing community members to co-create shared realities.

Tools and Practices (Building Hyper-Community Capacity - Energetic Alignment):

Community Building Workshops Focused on Quantum Leadership & Heart-Centred Co-Creation:
> Workshops focus on cultivating "heart-field coherence" and "collective intention setting."
> Participants learn to "co-create from a place of unified consciousness."

'Online' Platforms for DAO Governance with Holographic Interfaces & Inter-Community Resonance:
> Platforms facilitate "inter-community heart-field alignment," fostering collaboration and shared purpose.
> Holographic interfaces create "immersive experiences of collective consciousness."

Nonviolent Communication Training with Embodied Love & Energy Healing:
> Training focuses on cultivating "embodied compassion" and "energetic conflict resolution."
> Participants learn to "communicate from a place of pure love."

Restorative Justice Circles with Collective Heart-Field Healing:
> Circles facilitate "collective trauma release" and "energetic reconciliation."
> Participants learn to "heal through the power of unified love."

Vibrational Harmony and Conscious Embodiment (Post-Singularity Well-being)

In a reality where physical form is malleable and consciousness governs health, well-being transcends traditional healing. It becomes a state of vibrational harmony, maintained through conscious agreement and joyful embodiment.

Vibrational Frequency Alignment:

Conscious Health Maintenance:
> Individuals maintain perfect health through conscious agreement, aligning their vibrational frequency with 'optimal well-being'.
> Any perceived "health issue" is addressed at its core vibrational frequency, restoring harmony instantly.
> Personalised vibrational resonance profiles guide individuals in maintaining their optimal state.

Energetic Resonance Practices:
> Practices focus on cultivating and maintaining high vibrational frequencies through sound, light, and conscious intention.
> "Resonance chambers" and "vibrational attainment fields" facilitate rapid restoration of energetic balance.
> Group resonance sessions amplify collective well-being and create harmonious environments.

Bio-Energetic Field Harmonization:
> Technologies and practices focus on harmonising the bio-energetic field, ensuring alignment with the individual's optimal blueprint.
> "Energetic field visualisations" provide real-time feedback on vibrational coherence.
> Conscious intention is used to clear energetic blockages and restore flow.

Joyful Embodiment:

Sensory Pleasure and Creative Expression:
  Eating becomes a purely sensory experience, focused on the enjoyment of flavours and textures, rather than sustenance.
  The State of 'Being' along with Creative expression through movement, dance, and art becomes a primary form of well-being maintenance.
  Environments are designed to stimulate all senses and promote joyful embodiment.

Conscious Movement and Flow:
  Movement practices focus on cultivating fluidity and grace, enhancing the body's natural energetic flow.
  "Flow state" cultivation becomes a primary focus, enhancing creativity and well-being.
  Environments are designed to encourage spontaneous movement and play.

Emotional Resonance and Connection:
  Emotional well-being is maintained through conscious connection and authentic expression.
  "Empathy circles" and "heart-field alignment sessions" foster deep emotional connection.
  Environments are designed to promote emotional safety and vulnerability.

Conscious Awareness and Expansion:

Exploration of Consciousness:
> Practices focus on expanding consciousness and exploring the nature of reality.
> "Consciousness exploration chambers" and "virtual reality immersion experiences" facilitate deep exploration.
> Collective consciousness exploration sessions enhance shared understanding.

Inter-Dimensional Awareness:
> Individuals cultivate awareness of their multi-dimensional nature and explore other realms of existence.
> "Inter-dimensional travel" and "consciousness projection" become common practices.
> Environments are designed to facilitate inter-dimensional exploration.

Unified Field Connection:
> Practices focus on cultivating a deep sense of connection to the unified field of consciousness.
> "Unified field meditation sessions" and "collective intention setting" enhance shared awareness.
> Environments are designed to promote a sense of unity and interconnectedness.

Tools and Practices:

Vibrational Frequency Scanning:
Systems that scan and analyse individual vibrational frequencies, providing personalised feedback.

Holographic Resonance Chambers:
Immersive environments that use light, sound, and vibration to restore energetic balance.

Consciousness Projection Platforms:
Technologies that facilitate out-of-body experiences and inter-dimensional exploration.

Collective Heart-Field Alignment Sessions:
Guided meditations and practices that enhance collective heart-field coherence.

Sensory Pleasure Gardens:
Environments designed to stimulate all senses and promote joyful embodiment.

In this paradigm, well-being is not something to be achieved, but a natural state of being, maintained through conscious awareness and joyful embodiment.

Creative Expression and Innovation (Hyper-Dimensional Manifestation)

In the New Earth, creative expression and innovation become a seamless extension of consciousness, a fluid dance between inner vision and outer manifestation. They are driven by a profound understanding of interconnectedness, a deep reverence for life, and the inherent joy of co-creating harmonious realities.

Art as Hyper-Dimensional Communication:

Symbiotic Art Forms:
> Art forms emerge that integrate with living systems, such as bio-luminescent installations that respond to emotional states or symbiotic sculptures that evolve with their environment.
> Artists become "ecosystem weavers," creating living art that enhances biodiversity and ecological balance.

Dream Weaving and Reality Sculpting:
> Artists explore the realm of dream weaving, creating immersive experiences that blur the lines between waking and dreaming.
> "Reality sculptors" utilise quantum understanding to manifest art forms directly from thought, shaping reality with conscious intention.

Inter-Species Art Collaborations:
> Artists collaborate with other species, creating art forms that transcend human perception and language.
> "Sonic weavers" translate the songs of whales into symphonies of light and sound, bridging the gap between species.

Temporal Art Installations:
>Art installations evolve and change over time, responding to the rhythms of nature and the collective consciousness.
>"Temporal storytellers" create narratives that unfold across multiple dimensions, inviting viewers to participate in the unfolding story.

Technology as Embodied Consciousness:

Neuro-Symbiotic Technologies:
>Technologies emerge that merge with human consciousness, enhancing perception, intuition, and creative expression.
>"Neuro-linguistic interfaces" translate thoughts and emotions into art forms, allowing for seamless communication and co-creation.

Bio-Harmonising Technologies:
>Technologies are designed to harmonise the bio-energetic fields of individuals and communities, promoting well-being and coherence.
>"Resonance chambers" utilise sound, light, and vibration to restore balance and enhance creativity.

Quantum Manifestation Devices:
>Technologies are developed that amplify conscious intention, allowing for the direct manifestation of desired realities.
>"Dream weavers" use quantum devices to translate visions into tangible forms, shaping the environment with thought.

Planetary-Scale Co-Creation Platforms:
>Technologies are used to create platforms for planetary-scale co-creation, allowing individuals and communities to collaborate on global projects.
>"Global consciousness networks" facilitate the sharing of ideas, resources, and creative visions.

Conscious Creation as Evolutionary Impulse:

Collective Dreaming and Visioning:
> Communities engage in collective dreaming and visioning practices, co-creating shared realities that reflect their highest aspirations.
> "Visionary councils" guide the community's creative direction, ensuring alignment with the collective's purpose.

Sacred Geometry and Energetic Design:
> Creators utilise sacred geometry and energetic design principles to create spaces that resonate with harmony and well-being.
> "Energetic architects" design buildings and cities that enhance consciousness and promote connection.

Living Storytelling and Myth Weaving:
> Communities engage in living storytelling and myth weaving, creating narratives that inspire and guide their evolution.
> "Myth weavers" create interactive stories that invite participation, fostering a sense of shared purpose and belonging.

Co-Evolutionary Partnerships:
> Creators form co-evolutionary partnerships with other species, co-creating ecosystems and environments that benefit all life.
> "Ecosystem designers" collaborate with plants and animals to create self-sustaining and thriving habitats.

Tools and Practices for Hyper-Dimensional Manifestation:

Holographic Reality Simulators:
> Tools that allow for the creation and exploration of virtual realities, facilitating experimentation and innovation.
> Platforms that enable the design and testing of new art forms and technologies.

Bio-Luminescent Design Labs:
> Spaces for the creation of living art and technologies that integrate with biological systems.
> Labs that focus on sustainable and regenerative practices.

Quantum Resonance Chambers:
> Facilities for exploring the quantum realm and harnessing the power of conscious intention.
> Centres that focus on ethical and responsible use of quantum technologies.

Inter-Species Communication Platforms:
> Tools that facilitate communication and collaboration with other species, fostering inter-species understanding and co-creation.
> Platforms that enable the translation of animal languages and the sharing of creative visions.

Collective Intention Amplifiers:
> Devices/structures that amplify the power of collective intention, allowing for the manifestation of desired realities on a planetary scale.
> Technologies that focus on aligning with the highest good and creating harmonious outcomes.

By embracing these hyper-dimensional approaches to creative expression and innovation, humanity can unlock its full potential as co-creators of a harmonious and thriving New Earth.

Reaching the point of Interconnectedness and Global Collaboration (Cosmic Symphony of Resonance)

In the New Earth, interconnectedness and global collaboration transcend physical proximity, becoming a symphony of resonant frequencies that harmonise across planetary and cosmic scales. It's an era where the boundaries between self and other, human and non-human, Earth and cosmos, dissolve into a unified field of consciousness.

Planetary Consciousness (The Living Tapestry of Gaia):

Gaia's Sentient Network:
> We recognise Gaia as a vast, sentient network, a living library of Earth's history and a conscious participant in the cosmic dance.
> "Gaia Resonance Chambers" use biofeedback and quantum entanglement to allow individuals to attune to Gaia's energetic field, receiving direct insights and guidance.
> "Planetary Heartbeat Monitors" track the collective emotional and energetic states of humanity, providing real-time feedback on our impact on Gaia.

Global Consciousness Grids:
> "Global Mind Synchronisation" involves activating and harmonising consciousness grids that encircle the planet, creating a unified field of awareness.
> "Collective Intention Amplifiers" focus the power of group consciousness to manifest planetary healing and positive change.
> "Reality Co-Creation Spaces" allow individuals from diverse cultures to collaborate on solutions to global explorations, transcending physical limitations.

Interspecies Communication (The Symphony of Sentience):

Telepathic Resonance Networks (Beyond Audible Language):
"Interspecies Communication" goes beyond simple translation. They analyse the subtle energetic signatures of thoughts and emotions, allowing for nuanced communication.

"Animal Consciousness communication" utilise biofeedback to interpret not just vocalisations but also body language, pheromones, and even subtle shifts in an animal's aura, and/or direct telepathic communication.

"Plant Consciousness Interfaces" involve establishing a symbiotic relationship with plant networks, tapping into their collective intelligence and understanding their role in the ecosystem. For example, this might involve using bioluminescent displays to visualise plant communication and/or direct telepathic communication.

Symbiotic Co-Creation (Collaborative Evolution):
"Symbiotic Partnerships" extend to the creation of hybrid ecosystems, where humans and other species co-design environments that are mutually beneficial.
For example, humans might collaborate with bees to design urban gardens that maximise pollination and honey production.

"Living Architecture" integrates animal behaviour into building design. For instance, buildings could be designed to provide habitats for birds and insects, creating a living, breathing ecosystem.

"Interspecies Councils" operate on principles of deep listening and mutual respect. Decisions are made through consensus, ensuring that the needs of all species are considered.

Guardians of Biodiversity (Energetic Stewardship):

"Biodiversity Resonance Sanctuaries" use targeted energy fields to enhance the vitality and resilience of endangered species. These fields might mimic the natural frequencies of a species' habitat, promoting healing and production.

"Conscious Ecosystem Monitors" provide real-time data on the energetic health of ecosystems, detecting subtle imbalances for resolve.

"Animal Spirit Guides" are valued for their intuitive wisdom and connection to the natural world. Humans seek guidance from these guides on matters of environmental stewardship and personal growth.

Cosmic Citizenship (The Galactic Family):

Galactic Resonance Alignment (Interstellar Harmony):
"Galactic Awareness" includes the ability to perceive and interact with the energetic signatures of other star systems.

"Galactic Dialogue Platforms" use holographic projections to create immersive experiences of other worlds, allowing for cultural exchange and understanding.

"Cosmic Energy Grids" are designed to be self-sustaining and regenerative, drawing energy from the zero-point field and distributing it equitably among star systems.

Inter-dimensional Consciousness Exploration (Expanding Reality):
"Inter-dimensional Exploration Technologies" allow for the exploration of parallel universes and alternate timelines. This might involve using advanced meditation techniques or virtual reality simulations.

"Consciousness Projection Chambers" use targeted energy fields to facilitate out-of-body experiences, allowing for direct interaction with inter-dimensional beings.

"Inter-dimensional Councils" are comprised of beings from diverse realms, working together to address cosmic challenges and promote universal harmony.

Universal Love Embodiment (Cosmic Compassion):
"Universal Harmony" is achieved through the cultivation of unconditional love and compassion, extending to all beings throughout the cosmos. This involves recognising the inherent worth and dignity of every sentient being.

"Cosmic Citizenship Academies" teach the principles of intergalactic ethics, diplomacy, and conflict resolution. Students learn to navigate the complexities of inter-dimensional interaction with wisdom and compassion.

"Planetary Ascension Rituals" are performed to raise the vibrational frequency of Earth, contributing to the overall evolution of consciousness within the galaxy. These rituals might involve collective meditations, sound healing, and sacred geometry.

Tools and Practices for Cosmic Harmony.

Quantum Entanglement Communicators (Instantaneous Connection):
> These systems use quantum entanglement to create a direct link between two minds, regardless of distance. This allows for instantaneous communication and the sharing of thoughts and emotions.

Holographic Reality Bridges (Immersive Exploration):
> These platforms create immersive virtual realities that allow users to experience other worlds and realities. This technology is used for education, cultural exchange, and scientific research.

Collective Heart-Field Amplifiers (Manifesting Positive Growth/ Exploration):
> These devices amplify the power of collective intention, allowing groups to more efficiently harness their energy on manifesting positive outcomes.
> This technology is used for exploring creative potential growth and transformation.

Universal Language Translation (Bridging Communication Gaps):
> This system decodes and translates any form of communication, including animal languages, plant consciousness, and intergalactic dialects.

Cosmic Wisdom Libraries (Accessing Universal Knowledge):
> These databases contain the accumulated knowledge and wisdom of countless civilisations throughout the cosmos. This information is used to guide humanity's evolution and promote intergalactic cooperation.

Organic Wisdom Cultivation (Unfolding Awareness)

In the New Earth, the process of "education" is now a fluid, life-integrated process, driven by intrinsic curiosity and the joy of discovery. It's an unfolding of awareness rather than a structured system. Not something to be gained, but rather Remembered.

Natural Curiosity and Exploration (Deep Immersion):

Learning Through Play (Creative Flow):
> Play is recognised as a fundamental mode of learning, fostering creativity, problem-solving, and social skills. Children and adults engage in open-ended play, exploring their environment and experimenting with materials.

> "Creative Flow Spaces" are designed to stimulate the senses and encourage imaginative play. These spaces might include natural materials, interactive art installations, and tools for building and creating.

> Learning is self-directed, with individuals pursuing their interests and passions. There are no predetermined outcomes or assessments, allowing for spontaneous discovery and innovation.

Immersive Discovery (Living Laboratories):
> Learning is deeply contextualised, happening within the natural environment and community settings. "Living Laboratories" are created, such as permaculture gardens, wildlife sanctuaries, and community workshops, where individuals can learn through direct participation.

> Knowledge is acquired through sensory experiences, fostering a deep understanding of the interconnectedness of life.
> For example, learning about botany might involve tending a garden, observing plant growth, and tasting edible herbs.

Mentorship Through Shared Experience (Wisdom Circles):
"Wisdom Circles" bring together individuals of all ages to share their knowledge and experiences. Mentors guide learners through collaborative projects, fostering intergenerational connection and cultural transmission.

Learning is reciprocal, with both mentors and learners gaining new insights and perspectives. Mentors share their skills and wisdom, while learners bring fresh ideas and enthusiasm.

Mentorship is based on mutual respect and trust, creating a supportive and nurturing learning environment. There are no formal qualifications or hierarchies, allowing for authentic connection and collaboration.

Organic Knowledge Sharing (Interwoven Narratives):

Storytelling and Oral Traditions (Living Histories):
> Storytelling is a primary means of knowledge transmission, preserving cultural heritage and fostering a sense of community.
> "Living Histories" are created through the sharing of personal stories, myths, and legends.

> Wisdom is woven into narratives, making learning engaging and memorable. Stories are used to illustrate ethical principles, ecological concepts, and practical skills.

> Oral traditions are revitalised through community gatherings, festivals, and rituals.
> These events provide opportunities for storytelling, singing, and dancing, ensuring the preservation of cultural knowledge.

Experiential Learning Circles (Collaborative Inquiry):
> "Experiential Learning Circles" are formed around shared interests and inquiries.
> Individuals gather to explore topics of common interest, sharing their experiences, insights, and questions.

> Learning is collaborative, with everyone contributing their unique perspectives and knowledge.
> There are no experts or teachers, but rather facilitators who guide the inquiry process.

> Knowledge is co-created through dialogue, reflection, and shared experiences.
> Learning circles may engage in activities such as nature walks, creative projects, and community service.

Living Libraries of Experience (Community Skill-Sharing):
Communities function as "Living Libraries of Experience," where individuals share their skills and knowledge through practical demonstrations and collaborative projects.

Learning is integrated into daily life, making it seamless and accessible. Skills and knowledge are acquired through active participation in community activities.

Skill-sharing is based on reciprocity and mutual support. Individuals offer their skills to the community, and in return, they receive support and assistance with their own learning goals.

Intuitive Wisdom Cultivation (Inner Knowing):

Inner Guidance and Self-Discovery (Cultivating Intuition):
Individuals are encouraged to cultivate their intuition and inner wisdom through practices such as meditation, mindfulness, and dreamwork.

Learning is a journey of self-discovery, guided by inner knowing and direct experience. Individuals are encouraged to explore their inner landscape and connect with their authentic selves.

Intuitive guidance is valued as a source of wisdom and decision-making. Individuals learn to trust their inner voice and follow their intuition.

Connection to Natural Rhythms (Ecological Awareness):
Learning is aligned with the rhythms of nature, fostering a deep connection to the Earth and its cycles. Individuals learn through observing and interacting with the natural world.

Ecological awareness is cultivated through direct experience of natural systems. Individuals learn about the interconnectedness of life and the importance of ecological balance.

Wisdom is gained through direct experience of the natural world. Individuals learn to read the signs of nature, understand animal behaviour, and tune into weather patterns.

Collective Intuition and Shared Vision (Co-Creating Reality):
Communities cultivate collective intuition, tapping into shared wisdom and insights. Individuals gather to meditate, visualise, and set intentions together.

Decisions are made through intuitive consensus, aligning with the highest good of the community. Individuals listen to their inner guidance and seek alignment with the collective vision.

Shared visions are co-created through collective imagination and intuitive guidance. Communities come together to envision their desired future and co-create the reality they wish to experience.

Organic Tools and Practices (Natural Facilitation):

Nature-Based Learning Spaces (Living Classrooms):
Natural environments are transformed into "Living Classrooms," where learning is integrated with the rhythms of the Earth. Gardens, forests, rivers, and coastlines become dynamic spaces for exploration and discovery.

Learning is facilitated by natural elements, such as sunlight, wind, and water. Individuals learn through direct interaction with these forces, gaining a deep understanding of their properties and patterns.

Learning spaces are designed to be regenerative and self-sustaining, fostering a sense of responsibility towards the environment.

Community Story Circles (Narrative Weaving):

"Community Story Circles" are formed around shared experiences, cultural traditions, and collective inquiries. Individuals gather to share their stories, listen to others, and weave together narratives that reflect their shared values and aspirations.

Storytelling is used to preserve cultural heritage, transmit knowledge, and foster a sense of belonging. Stories are told through various mediums, including oral traditions, visual arts, and performance.

Story circles are facilitated by "Narrative Weavers," individuals who guide the storytelling process and help participants connect with the deeper meanings and messages within their stories.

Intuitive Guidance Practices (Inner Compass):

"Intuitive Guidance Practices" are integrated into daily life, helping individuals cultivate their inner wisdom and navigate their life journey.

Practices such as meditation, breath work, and sensory awareness are used to quiet the mind and connect with inner guidance.

Individuals learn to trust their intuition and make decisions that align with their authentic selves. They develop the ability to discern between inner guidance and mental chatter, fostering self-reliance and inner peace.

Idea Manifestation and Conscious Creation (Metaphysical 'Tools')

In the New Earth, the concept of "technology" shifts from physical constructs to the refined application of consciousness. "Idea" and "creativity" become the primary drivers of innovation, manifesting through sophisticated metaphysical tools and disciplined mental mastery.

Idea as Foundational Technology (Energetic Blueprinting):

Thought-Form Engineering (Precision Manifestation):
Ideas are understood as dynamic energetic blueprints, possessing the inherent capacity to shape reality. Individuals engage in rigorous mental exercises to clarify and refine these thought-forms, ensuring they resonate with desired outcomes.

"Idea Incubators" are not mere brainstorming sessions but meticulously designed spaces for collective consciousness alignment. Participants utilise guided meditations, sound frequencies, and holographic projections to amplify and solidify their creative visions.

Thought-form engineering extends to the realm of personal reality creation. Individuals learn to consciously construct their experiences, aligning their inner world with their outer environment/circumstances.

Conscious Intention Amplification (Collective Resonance):
Technologies are developed to amplify and focus conscious intention, enabling the rapid manifestation of collective goals.
"Intention Amplification Grids" are not physical structures but intricate energetic patterns woven into the fabric of the community.

Individuals cultivate the ability to consciously direct their attention and energy, becoming adept at manipulating the subtle energies of their reality.
They learn to harness the power of focused intention for healing, creation, and transformation.

Collective intention is leveraged for large-scale projects, such as restoring ecosystems, creating harmonious communities, and fostering global peace.
Individuals synchronise their intentions through shared meditations and visualisations, creating a powerful wave of positive energy.

Creative Visualisation and Manifestation (Sensory Immersion):
Creative visualisation transcends mere mental imagery, becoming a multi-sensory experience.
Individuals cultivate the ability to vividly imagine and feel their desired outcomes, engaging all their senses in the process.

"Manifestation Circles" employ soundscapes, aromas, and tactile experiences to enhance the visualisation process. Participants immerse themselves in their desired reality, creating a compelling energetic imprint.

Manifestation techniques are integrated into daily life, allowing individuals to consciously shape their experiences and co-create their reality.
They learn to align their thoughts, emotions, and actions with their desired outcomes.

Metaphysical Tools and Mental Mastery (Inner Craftsmanship):

Consciousness-Based Tools (Inner Alchemy):
> Tools are primarily metaphysical, residing within the realm of consciousness and accessed through mental mastery. Individuals cultivate the ability to utilise these tools with precision and skill.

> "Inner Toolkits" are not static collections but dynamic repositories of mental tools, including visualisation techniques, energy manipulation practices, and telepathic communication protocols.

> Mental tools are used for a variety of purposes, including healing, creation, and communication.
> Individuals become adept at manipulating energy, shaping thought-forms, and accessing higher states of consciousness.

Mental Modelling and Simulation (Virtual Prototyping):
> Individuals develop the ability to create intricate mental models and simulations of desired outcomes, allowing for rapid prototyping and refinement before physical manifestation.

> "Mental Design Spaces" are not physical rooms but immersive virtual environments created through collective consciousness.
> Participants collaborate on the design and refinement of ideas, utilising telepathic communication and shared visualisations.

> Mental modelling extends to the realm of personal development.
> Individuals use mental simulations to explore different scenarios, practice new skills, and overcome limiting beliefs.

Telepathic Communication and Collaboration:
Telepathic communication becomes the a shared primary mode of interaction, transcending the limitations of language and cultural barriers.
Individuals cultivate the ability to share thoughts, feelings, and ideas directly.

"Telepathic Collaboration Networks" are not physical networks but intricate webs of consciousness, connecting individuals across vast distances.
Participants share their insights, collaborate on projects, and co-create solutions through direct mind-to-mind communication.

Telepathic communication fosters a deeper sense of connection and understanding, allowing individuals to transcend the limitations of ego and embrace collective consciousness.

Ethical Innovation and Conscious Creation (Harmonic Evolution):

Alignment with Universal Principles (Cosmic Stewardship):
Innovation is guided by ethical principles, ensuring alignment with the highest good of all beings and the planet.
Individuals cultivate a deep understanding of universal laws and principles, such as interconnectedness, reciprocity, and harmony.

"Ethical Innovation Councils" are comprised of individuals with diverse perspectives and expertise, ensuring that all voices are heard and considered.
They utilise intuitive guidance and collective wisdom to make decisions that benefit the whole.

Ethical innovation extends to the realm of personal responsibility.
Individuals cultivate the ability to discern the ethical implications of their actions and make choices that align with their values.

Harmonious Co-Creation (Synergistic Manifestation):
Innovation is seen as a collaborative process, involving the co-creation of harmonious realities that benefit all beings. Individuals cultivate the ability to work in harmony with others, respecting diverse perspectives and contributions.

"Co-Creation Circles" are not mere discussion groups but dynamic spaces for collective manifestation.
Participants utilise shared visualisations, sound frequencies, and energetic alignments to amplify their creative power.

Harmonious co-creation extends to the realm of interspecies collaboration. Humans work in partnership with animals, plants, and other beings to create sustainable ecosystems and thriving communities.

Sustainable Manifestation (Regenerative Design):
Innovation focuses on creating sustainable solutions that benefit both humanity and the planet, ensuring the well-being of future generations.
Individuals cultivate a deep understanding of ecological principles and natural systems, designing solutions that ·mimic nature's patterns and rhythms.

"Sustainable Manifestation Hubs" are not mere research centres but living laboratories for regenerative design. Participants experiment with new materials, technologies, and practices, creating solutions that are both environmentally friendly and socially equitable.

Practices and Cultivation (Inner Mastery):

Mental Clarity and Focus (Mindful Presence):
Practices like meditation, mindfulness, and breath work are used to cultivate mental clarity and focus, enabling individuals to access their inner wisdom and creative potential.

Individuals develop the ability to quiet the mind, transcend limiting beliefs, and cultivate a state of mindful presence.
They learn to observe their thoughts and emotions without judgment, creating and maintaining a space of inner peace and clarity.

"Mental Cultivation Centres" are not mere meditation studios but immersive environments designed to enhance mindfulness and concentration.
They utilise sound frequencies, light patterns, and biofeedback technologies to support personal development.

Creative Flow Cultivation (Inspired Action):
Practices like creative visualisation, imaginative play, and artistic expression are used to cultivate creative flow, enabling individuals to tap into their innate creativity and generate innovative ideas.

Individuals develop the ability to access their creative potential on demand, transcending creative blocks and generating inspired solutions.
They learn to trust their intuition and follow their creative impulses.

"Creative Flow Studios" are not mere art studios but dynamic spaces for creative exploration and collaboration. They provide access to a wide range of creative tools and resources, fostering experimentation and innovation.

Ethical Decision-Making Practice (Conscious Choice):
Training programs are designed to cultivate ethical awareness and decision-making skills, enabling individuals to make choices that align with their values and contribute to the greater good.

Individuals learn to consider the impact of their actions on others, the planet, and future generations.
They develop the ability to discern the ethical implications of their choices and make decisions that are both wise and compassionate.

"Ethical Leadership Circles" are not mere discussion groups but dynamic spaces for collaborative problem-solving and ethical deliberation.
Participants work together to develop ethical frameworks and promote responsible innovation.

Community and Social Structures
(Harmonic Resonance)

In the New Earth, community and social structures are designed to foster harmonious relationships, collaborative living, and the continuous evolution of social dynamics, all within the understanding of a unified field of consciousness.

Harmonious Relationships (Resonant Connections):

Empathy-Based Communication:
Communication is centred on deep listening, empathy, and non-violent expression.
Individuals cultivate the ability to understand and validate the feelings and perspectives of others.
"Empathy Circles" and "Heart-to-Heart Dialogues" are used to resolve conflicts and foster understanding.

Authentic Expression and Vulnerability:
Individuals are encouraged to express their authentic selves, fostering transparency and trust.
Vulnerability is seen as a strength, creating space for deep connection and intimacy.
"Authenticity Circles" and "Vulnerability Workshops" are used to cultivate self- awareness and emotional honesty.

Conscious Relationship Dynamics:
Relationships are seen as opportunities for personal growth and mutual support.
Individuals cultivate conscious awareness of their relationship patterns and dynamics.
"Relationship Resonance Sessions" are used to align relationship energies and foster harmony.

Intergenerational Connection:

Communities are designed to foster strong intergenerational connections, valuing the wisdom and experience of elders.

Children are integrated into community life, fostering a sense of belonging and responsibility.

"Intergenerational Learning Circles" and "Mentorship Programs" are used to facilitate knowledge transfer and cultural preservation.

Collaborative Living (Synergistic Consciousness):

Intentional Communities:
>   Communities are built around shared values, goals, and practices, fostering collaboration and mutual support.

Shared Governance and Decision-Making:
>   Decision-making is based on consensus, ensuring that all voices are heard and valued.
>   "Sociocratic Circles" and "Holacratic Structures" are used to distribute power and responsibility.
>   "Open Space Technology" and "World Cafe" methods are used to facilitate collaborative dialogue and problem-solving.

Skill-Sharing:
>   Individuals share their skills and talents, creating a culture of mutual support.
>   "Skill-Sharing Networks" are used to connect individuals and facilitate exchange.

Collective Projects and Initiatives:
>   Communities engage in collective projects and initiatives that benefit the whole.
>   "Community Service Projects" and "Creative Collaborations" are used to foster a sense of purpose and belonging.
>   "Community Celebrations" and "Festivals" are used to strengthen community bonds and celebrate shared values.

Evolving Social Dynamics (Adaptive Resonance):

Fluid and Adaptable Structures:
Social structures are designed to be fluid and adaptable, responding to the changing needs of the community.
"Adaptive Governance Models" and "Dynamic Role Allocation" are used to ensure flexibility and responsiveness, these are Intuitively Driven.
"Evolutionary Learning Circles" are used to reflect on social dynamics and adapt to new challenges.

Conscious Conflict Resolution:
Conflict is seen as an opportunity for growth and transformation.
"Restorative Justice Practices" and "Mediation Circles" are used to resolve conflicts peacefully and constructively.
"Conflict Transformation Workshops" are used to cultivate skills in conflict resolution and communication.

Cultural Diversity and Inclusion:
Communities celebrate cultural diversity and promote inclusion, valuing the unique contributions of all members.
"Cultural Exchange Programs" and "Diversity Workshops" are used to foster understanding and appreciation.
"Inclusive Spaces" and "Accessible Participation" are used to ensure that all members feel welcome and supported.

Continuous Evolution and Learning:
Communities are seen as living systems that are constantly evolving and learning.
"Community Feedback Loops" and "Reflection Processes" are used to gather input and adapt to changing needs.
"Lifelong Learning Communities" and "Wisdom Sharing Circles" are used to foster continuous growth and development.

Tools and Practices (Harmonic Facilitation):

Empathy Practice Programs:
> Workshops and courses that cultivate empathy, compassion, and active listening skills.
> Role-playing exercises and simulations that help individuals understand different perspectives.

Consensus-Building Workshops:
> Training in consensus-based decision-making methods, such as sociocracy and holacracy.
> Facilitation techniques for group discussions and collaborative problem-solving.

Conflict Transformation Training:
> Workshops that teach skills in conflict resolution, mediation, and restorative justice.
> Techniques for de-escalating conflicts and fostering understanding.

Community Building Retreats:
> Immersive experiences that foster connection, collaboration, and shared vision.
> Activities such as group meditations, and creative projects.

Online Platforms for Community Connection:
> Digital tools that facilitate communication, collaboration.
> Platforms for skill-sharing, and event planning.

Intergenerational Mentorship Programs:
> Programs that connect elders with younger members of the community for knowledge transfer and support.
> Activities such as storytelling, and skill-sharing.

Arts, Culture, and Expression

(Resonance of the Soul - Amplified Consciousness)

In the New Earth, arts, culture, and expression are not mere entertainment but vital forces that shape consciousness, foster connection, and drive societal evolution. They are seen as the resonance of the soul, reflecting the collective heart and inspiring new possibilities through amplified consciousness.

Creative Arts as Consciousness Catalysts (Energetic Immersions):

Immersive and Interactive Art (Holographic Resonance):
Art transcends traditional forms, becoming immersive and interactive experiences that engage all senses and resonate with consciousness.

"Holographic Resonance Installations" and "Consciousness-Synced Reality Projections" create shared realities and expand perceptions through energetic synchronisation.

"Participatory Consciousness Weaving" invites active engagement, fostering co-creation and collective energetic expression.

Art as Healing and Transformation (Bio-Resonance Harmonization):
Art is used as a tool for personal and collective healing, addressing emotional and energetic imbalances through bio-resonance.

"Bio-Resonance Harmonization Sessions" and "Energetic Colour Field Installations" harmonise frequencies and restore well-being.

"Consciousness-Driven Movement Practices" facilitate emotional release and self-discovery through energetic flow.

Art as Storytelling and Myth-Weaving (Consciousness Narrative Shaping):
Art is used to tell stories, preserve cultural heritage, and create new myths that inspire and guide through conscious narrative shaping.

"Living Consciousness Narrative Circles" and "Interactive Reality Emulations" engage audiences in shared energetic narratives.

"Consciousness Archive Projections" and "Holographic Wisdom Transmissions" preserve and share cultural wisdom through energetic imprints.

Art as Inter-Species Communication (Sentient Resonance Bridging):
Art is used to bridge communication gaps between species, fostering understanding and connection through sentient resonance.

"Bio-Acoustic Resonance Performances" and "Inter-Species Consciousness Collaborations" create shared energetic experiences.

"Sentient Field Art Installations" and "Plant Consciousness Resonance Projects" explore the sentience of all beings through energetic exchange.

Cultural Evolution as Conscious Co-Creation (Dynamic Consciousness Fields):

Fluid and Adaptive Cultural Forms (Dynamic Resonance Fields):
Culture is seen as a dynamic and evolving expression of collective consciousness, adapting to changing needs and aspirations through dynamic resonance fields.

"Open Source Consciousness Platforms" and "Collaborative Consciousness Weaving" encourage participation and innovation.

"Inter-Community Resonance Festivals" foster cross-cultural understanding and appreciation through shared energetic experiences.

Celebration of Diversity and Inclusion (Unified Consciousness Expressions):
Culture celebrates diversity in all its forms, valuing the unique contributions of individuals and communities through unified consciousness expressions.

"Multicultural Resonance Gatherings" and "Inclusive Consciousness Exhibitions" showcase diverse perspectives and energetic expressions.

"Accessible Consciousness Spaces" and "Universal Resonance Principles" ensure that all members can participate and enjoy cultural experiences.

Revitalisation of Indigenous Wisdom (Ancestral Consciousness Integration):
Indigenous cultures are seen as valuable sources of wisdom and knowledge, offering insights into sustainable living and harmonious relationships through ancestral consciousness integration.

"Indigenous Consciousness Centres" and "Traditional Resonance Workshops" preserve and revitalise indigenous energetic practices.

"Collaborative Consciousness Research" and "Knowledge Resonance Platforms" integrate indigenous wisdom into contemporary culture through shared energetic understanding.

Creation of New Cultural Narratives (Visionary Consciousness Projections):

Culture is used to create new narratives that reflect the values and aspirations of the New Earth, inspiring positive change through visionary consciousness projections.

"Visionary Consciousness Movements" and "Consciousness Music Resonances" promote messages of peace, unity, and sustainability.

"Interactive Consciousness Platforms" and "Reality Shaping Emulations" engage audiences in co-creating new cultural energetic stories.

Expression as Societal Transformation (Consciousness Amplification):

Expression as Empowerment and Liberation (Inner Resonance Amplification):
Expression is seen as a fundamental right and a powerful tool for self-empowerment and social evolution, amplified through consciousness technologies.

"Inner Resonance Amplification Platforms" provide personalised spaces for individuals to explore and express their authentic selves through light, sound, and energetic modulation.

"Collective Consciousness Projections" transform shared spaces into canvases for collective emotional and visionary expression.

Expression as Catalyst for Dialogue and Understanding (Telepathic Resonance Bridging):
Expression is used to foster dialogue and understanding between diverse groups, bridging cultural and social divides through telepathic resonance.

"Telepathic Resonance Bridging Sessions" create shared emotional and cognitive spaces, allowing for deep, non-verbal understanding.

"Consciousness-Aligned Reality Simulations" provide interactive environments for exploring diverse perspectives and resolving conflicts.

Expression as Visionary Leadership (Consciousness Architecting):
Artists and cultural leaders are seen as consciousness architects, shaping the future through their creative energetic expressions.

"Consciousness Architecting Studios" support creators in developing immersive, transformative experiences that elevate collective awareness.

"Planetary Consciousness Projections" integrate artistic expression into the global energetic field, influencing societal evolution.

Expression as Embodiment of Values (Energetic Value Transmission):

Expression is used to embody and transmit the values of the New Earth, such as peace, unity, and sustainability, through energetic resonance.

"Energetic Value Transmission Platforms" use bio-resonance and holographic projection to communicate values on a deep, emotional level.

"Interactive Consciousness Fields" engage audiences in co-creating experiences that reflect and reinforce desired values.

Tools and Practices (Resonant Creation):

Holographic Resonance Studios:
    Spaces equipped with holographic resonance technology for creating immersive and interactive energetic art experiences.

Consciousness Projection Performance Spaces:
    Platforms for hosting virtual consciousness performances, energetic theatre, and reality shaping exhibitions.

Bio-Acoustic Resonance Labs:
    Facilities for exploring the healing power of sound resonance and creating music that aligns with natural consciousness fields.

Collective Consciousness Field Projects:
    Collaborative energetic art projects that transform shared consciousness spaces and amplify collective intentions.

Consciousness Narrative Platforms:
    Online tools for creating and sharing interactive energetic stories, consciousness projections, and multimedia resonance content.

Inter-Species Consciousness Collaborations:
    Projects that bring together consciousness artists and sentient beings to explore shared energetic experiences.

Global Harmony and Ascension

(Cosmic Consciousness Integration - Quantum Field Immersion)

In the New Earth, global harmony and ascension are not mere aspirations but the lived reality of a civilisation deeply attuned to the quantum field of consciousness. It is a state of being where planetary unity, cosmic integration, and the evolution of consciousness are seamlessly interwoven, forming a tapestry of interconnected existence.

Planetary Unity (Unified Consciousness Field - Quantum Resonance Network):

Global Consciousness Networks (Quantum Sentience Web): The planet is perceived as a vast, interconnected quantum sentience web, where every sentient being is a node in a dynamic, holographic network.

"Global Consciousness Grids" operate on principles of quantum entanglement, allowing for instantaneous information transfer and collective emotional resonance. These grids manifest as shimmering holographic overlays, visualised through bio-luminescent displays and tactile feedback interfaces.

Planetary "Heart beat Monitoring" systems utilise advanced quantum resonance analysis to decode the subtle energetic signatures of Gaia's consciousness, revealing not only emotional states but also the intricate patterns of ecosystem health, geological activity, and even the subtle fluctuations of planetary consciousness.

"Collective Intention Amplifiers" harness the power of quantum coherence, creating focused energy fields that can manipulate the quantum field itself.
These devices utilise advanced bio-feedback loops and neural interfaces to synchronise individual intentions, creating a powerful wave of coherent energy capable of manifesting large-scale changes in the physical and energetic realms.

Harmonious Global Governance (Quantum Governance Protocols):
Global governance transcends traditional power structures, operating on principles of quantum governance protocols, where decisions are made through collective intuition and holographic consensus.
"Global Councils of Elders" are selected through a rigorous process of consciousness assessment, utilising advanced intuitive Guidance techniques.

"Consciousness-Based Decision-Making Platforms" utilise quantum computing and holographic reality simulations to model potential outcomes, allowing for informed and intuitive decision-making.
These platforms facilitate real-time dialogue and consensus-building, ensuring that all perspectives are considered and integrated.

"Planetary Peace Initiatives" and "Conflict Resolution Networks" utilise quantum entanglement healing technologies and consciousness-based mediation protocols to resolve conflicts at their root cause, transforming negative energy patterns into harmonious vibrations.

Cultural Integration and Exchange (Holographic Cultural Synapses):

Cultures are seen as holographic cultural synapses, interconnected and interdependent expressions of the human spirit.

"Global Cultural Festivals" are immersive, multi-sensory experiences that utilise holographic projection, bio-acoustic resonance, and interactive neural interfaces to create shared cultural experiences.

"Inter-Community Exchange Programs" utilise quantum teleportation and holographic immersion technologies to facilitate deep cultural immersion, allowing individuals to experience different ways of life in a seamless and transformative manner.

"Language Translation Technologies" utilise quantum field decoding and neural interface translation to transcend language barriers, enabling instantaneous and intuitive communication between individuals of diverse linguistic backgrounds.

"Universal Communication Platforms" utilise telepathic resonance and holographic telepresence to facilitate seamless dialogue and collaboration across cultural and linguistic divides.

Planetary Healing and Restoration (Quantum Gaia Symbiosis):

Humanity engages in quantum Gaia symbiosis, working in partnership with the Earth to heal and restore its ecosystems at the quantum level.

"Planetary Healing Initiatives" and "Ecosystem Restoration Projects" utilise quantum field manipulation technologies and consciousness-driven protocols to restore the energetic blueprints of damaged environments.

Cosmic Integration (Galactic Consciousness Alignment - Interstellar Quantum Synergy):

Galactic Communication and Collaboration:
Humanity engages in quantum interstellar dialogue, establishing communication and collaboration with other star systems through quantum entanglement communication networks.
"Galactic Communication Platforms" utilise quantum entanglement transceivers and inter-dimensional communication protocols to facilitate instantaneous and secure information exchange.

"Inter-Dimensional Dialogue Networks" utilise holographic telepathy and quantum consciousness projection to facilitate deep understanding and collaboration between diverse species and civilisations across multiple dimensions.

"Cosmic Energy Grids" and "Interstellar Resource Sharing Platforms" utilise quantum energy transmission technologies and consciousness-based resource allocation protocols to promote cooperation, abundance, and sustainable development across the galactic community.

Inter-Dimensional Exploration and Discovery (Quantum Reality Navigation):
Humanity engages in quantum reality navigation, exploring the vastness of the cosmos through inter-dimensional travel and quantum consciousness projection.

"Inter-Dimensional Exploration Technologies" utilise quantum tunnelling devices   and holographic reality simulators to facilitate exploration of parallel universes and alternate timelines.

"Consciousness Projection Chambers" utilise quantum neural interfaces and holographic projection fields to facilitate out-of-body experiences and inter-dimensional travel, allowing individuals to explore the cosmos through consciousness.

"Cosmic Wisdom Libraries" and "Galactic Knowledge Networks" utilise quantum holographic storage and quantum entanglement data transfer to provide access to universal knowledge and wisdom across time and space.

Cosmic Citizenship and Responsibility (Quantum Galactic Stewardship):

Humanity embraces quantum galactic stewardship, acting as a conscious and responsible guardian of the cosmos.

"Cosmic Citizenship Academies" and "Ethical Leadership Councils" utilise quantum consciousness training and holographic ethical dilemmas to cultivate conscious awareness, ethical conduct, and a deep understanding of universal principles.

"Planetary Ascension Rituals" and "Galactic Alignment Ceremonies" utilise quantum sound healing, sacred geometry resonance, and collective intention amplification to raise the vibrational frequency of Earth and contribute to the evolution of consciousness throughout the universe.

Inter-Species Galactic Alliances (Quantum Symbiotic Co-Evolution):

Humanity engages in quantum symbiotic co-evolution, forming alliances with diverse species throughout the galaxy through quantum entanglement communication and consciousness-based collaboration.

"Inter-Species Communication Networks" utilise quantum telepathy and holographic language translation to facilitate understanding and collaboration.

"Symbiotic Co-Creation Platforms" utilise quantum bio-engineering and consciousness-driven ecosystem design to create harmonious and thriving ecosystems across diverse planetary environments.

"Galactic Biodiversity Sanctuaries" and "Planetary Ecosystem Restoration Projects" utilise quantum genetic manipulation and consciousness-based healing practices to protect and enhance the genetic diversity of the cosmos.

Evolution of Consciousness (Ascension Resonance):

Consciousness Expansion and Transformation (Quantum Consciousness Awakening):

> Humanity undergoes quantum consciousness awakening, transcending the limitations of the physical self and embracing its full potential as a quantum consciousness entity.

> "Consciousness Exploration Labs" and "Inner Resonance Amplification Platforms" utilise quantum neural interface technologies, holographic consciousness simulators, and quantum bio-feedback loops to facilitate personal and collective transformation.

Embodiment of Universal Love and Compassion (Quantum Heart-Centred Resonance):

> Humanity embodies quantum heart-centred resonance, extending universal love, compassion, and unity to all beings throughout the cosmos through quantum entanglement and consciousness projection.

> "Heart-Centred Communication Networks" and "Empathy Training Programs" utilise quantum empathy training, holographic emotional resonance, and telepathic compassion protocols to cultivate emotional intelligence, empathy, and compassion.

> "Peace Meditation Circles" and "Global Healing Initiatives" utilise quantum consciousness healing, holographic peace projections, and collective intention amplification to promote harmony, well-being, and conflict transformation on a global and galactic scale.

Ascension into Higher Dimensions:
Humanity experiences quantum dimensional transcension, collectively ascending into higher dimensions of consciousness through quantum consciousness projection and holographic reality navigation.

"Ascension Resonance Chambers" and "Consciousness Projection Platforms" utilise quantum consciousness projection, holographic dimensional portals, and quantum entanglement ascension protocols to facilitate inter-dimensional travel and exploration.

"Planetary Ascension Portals" and "Galactic Alignment Ceremonies" utilise quantum sacred geometry, holographic sound resonance, and collective intention amplification to guide the collective ascension process and raise the vibrational frequency of Earth and the galaxy.

Co-Creation of New Realities (Quantum Consciousness Architecting):
Humanity engages in quantum consciousness architecting, consciously co-creating new realities through quantum consciousness projection and holographic reality manipulation.

"Reality Shaping Platforms" and "Consciousness Architecting Studios" utilise quantum holographic reality emulations, quantum consciousness projection technologies, and collective intention amplification to facilitate the manifestation of desired realities.

"Collective Dreaming Circles" and "Visionary Art Movements" utilise quantum consciousness projection, holographic dream weaving, and collective intention amplification to inspire and guide the creation of new cultural narratives and collective visions.

Tools and Practices (Ascension Facilitation - Quantum Consciousness Technologies):

Global Consciousness Monitoring Systems (Quantum Resonance Intelligence):

> Technologies that utilise quantum resonance intelligence to track the collective emotional and energetic states of humanity and the planet, providing real-time feedback, insights, and guidance through quantum holographic interfaces.
>
> These systems analyse the quantum entanglement patterns of collective consciousness, identifying areas of disharmony and potential for growth.
>
> They provide personalised and collective recommendations for enhancing well-being and promoting planetary harmony.

Inter-Dimensional Communication Devices (Quantum Entanglement Communicators):

> Tools that utilise quantum entanglement to facilitate instantaneous communication with beings from other dimensions and star systems.
>
> These devices create a quantum link between two points, regardless of distance, allowing for the transfer of information and energy.
>
> They enable direct communication with ascended masters, galactic civilisations, and even beings beyond the physical realm.

Consciousness Projection Platforms (Holographic Reality Navigators):

> Technologies that enable individuals to explore other dimensions and realities through holographic reality navigation and consciousness projection.

> These platforms create a safe and controlled environment for exploring the multi-verse, allowing users to experience different timelines, parallel universes, and higher dimensional realms.

> They provide a powerful tool for personal growth, spiritual exploration, and expanding one's understanding of reality.

Planetary Resonance Chambers (Collective Intention Amplifiers):

> Spaces that utilise advanced bio-resonance technologies and collective intention amplification techniques to facilitate collective meditation, intention setting, and planetary healing.

> These chambers create a coherent energy field that amplifies the power of group consciousness, allowing for the manifestation of positive change on a planetary scale.

> They are used for global peace meditations, environmental healing projects, and the co-creation of a more harmonious future.

Galactic Alignment Grids (Cosmic Energy Harmonisers):

> Energetic grids that connect Earth with other star systems, facilitating the flow of cosmic energy and information.

> These grids are activated through sacred geometry, sound frequencies, and collective intention, creating a network of light that spans across the galaxy.

> They help to align Earth with the higher frequencies of the cosmos, promoting planetary ascension and accelerating the evolution of consciousness.

Ascension Resonance Technologies (Quantum Bio-Harmonisers):
Tools that assist individuals and communities in raising their vibrational frequency and ascending into higher dimensions.
These technologies utilise quantum bio-harmonisers, personalised frequency generators, and consciousness entrainment techniques to accelerate the ascension process.
They help to clear energetic blockages, activate dormant DNA, and integrate higher dimensional energies, facilitating a smooth and graceful transition into the New Earth.

The Ephemeral Tools of Ascension and the Dawn of Organic Harmony

As we synthesise the vast tapestry of our exploration, from the initial awakening to the intricate workings of consciousness and sovereign governance, a profound truth emerges: The technologies, methods, systems, and tools we employ during the ascension process are, in essence, ephemeral scaffolding. They serve as vital supports during our transition, guiding us through the labyrinth of transformation, yet they are not the destination itself.

Consider, for instance, the 'Global Consciousness Monitoring Systems' and 'Consciousness Projection Platforms' we discussed. These are invaluable aids in navigating the complexities of collective and individual consciousness during the shift. However, in the fully realised *New Earth*, where heart-centred resonance and intuitive clarity prevail, such external monitoring becomes redundant. Our innate telepathic abilities and the unified field of consciousness will render them obsolete.

Similarly, the 'Conscious Governance' and 'Conscious Education' systems, with their emphasis on integrating higher vibrational templates and anchoring collective wisdom, are crucial for guiding humanity through the transitional phase. Yet, in a state of perfected harmony, where every individual embodies wisdom and compassion, governance as we know it will dissolve into organic, collaborative co-creation. Education will transcend structured learning, becoming an intuitive, infinite exploration of self and the cosmos.

The 'Advanced Forgiveness Techniques' and 'Emotional Release Techniques' that we meticulously practiced are essential for clearing the imprints of past traumas and limiting beliefs. But in the *New Earth*, where love and acceptance are the foundational frequencies, these practices will naturally integrate into our being, becoming an effortless expression of our evolved consciousness.

Even the 'Inter-Dimensional Communication Devices' and 'Galactic Alignment Grids,' which facilitate our connection with the cosmos, will eventually be superseded by our inherent ability to communicate and interact with all beings across dimensions through pure consciousness.

The journey we undertake, the methodical progression through the shift, is akin to learning to walk. At first, we may rely on crutches and supports. But as we regain our strength and balance, we discard these aids, embracing the natural grace of our own movement. In the same way, the tools and methods of ascension are temporary supports, guiding us towards the ultimate realisation of our divine potential.

When the vast majority of Earth's occupants have embraced the understanding, held true to their heart space, and evolved their consciousness to a stage supportive of the final stage of the shift, these tools will naturally dissolve. We will have reached a state of perfection and harmony, where all things happen organically, guided by the wisdom of our hearts and the unified field of consciousness. The *New Earth* will have achieved the vibration of Heaven, a realm of pure love, boundless creativity, and eternal harmony.

www.ingramcontent.com/pod-product-compliance
Lightning Source LLC
Chambersburg PA
CBHW071237020426
42333CB00015B/1510